Breaking New Ground

Frontispiece: "What We Have We Own." The cover of the Manitoba Engines Limited, Brandon, catalogue, 1915. (Ag.12. II. B. 48, SABS)

Breaking New Ground

A Century of Farm Equipment Manufacturing on the Canadian Prairies

DONALD G. WETHERELL
with ELISE A. CORBET

FIFTH HOUSE PUBLISHERS

in association with
Alberta Community Development

Cover design by John Luckhurst/GDL
Cover illustration: detail from the cover of the Manitoba Engines Limited, Brandon, catalogue, 1915, from the collection of the Saskatchewan Archives Board, Ag.12.II.B.48, Manitoba Engines Limited

Printed and bound in Canada
93 94 95 96 97 / 5 4 3 2 1

The publisher gratefully acknowledges the assistance of the Saskatchewan Arts Board, Communications Canada, and The Canada Council.

Canadian Cataloguing in Publication Data

Wetherell, Donald Grant, 1949–

Breaking new ground

Includes bibliographical references and index.
ISBN 1–895618–23–1

1. Agricultural implements - Trade and manufacture - Prairie Provinces - History. 2. Agricultural innovations - Prairie Provinces - History.
I. Corbet, Elise A. II.Title.

HD9486.C23P68 1993 338.4'7631'0971209 C93–098107–3

FIFTH HOUSE LTD.
620 Duchess Street
Saskatoon, SK
S7K 0R1

Contents

Acknowledgements

In 1990, the Prairie Implement Manufacturers Association (PIMA) celebrated its twentieth anniversary. To recognize this event, the association commissioned a history of farm equipment manufacturing on the Prairies. Support was also secured from all three provincial governments.

Many individuals have contributed to the study. Ron Zimmer, Ivan Thue, and Brett Filson have administered the project on behalf of PIMA. The former general manager of PIMA, Ronald Zimmer, was instrumental in initiating the project, and Ivan Thue also generously shared with us his knowledge of the industry. Dr. Carl Betke, and his successor, Dr. Michael Payne, Historic Sites Service, Alberta Community Development, administered the project on behalf of the province of Alberta.

While too numerous to name, others who shared their expertise include all those who were interviewed for the project and those members of PIMA who completed our questionnaire on the history of their companies. So, too, the staff of archives and libraries throughout the Prairies and in Ottawa provided crucial assistance. Patricia Myers of the Reynolds Alberta Museum and Dr. Ken Domier of Agricultural Engineering, University of Alberta, read and made comments on the first draft of the manuscript. Marshall Eliason of the Alberta Department of Agriculture also provided comments and assistance in locating photographs. A special note of thanks is due to Irene Kmet. She provided comment on the text and edited the various drafts.

PIMA has generously undertaken to place a copy of the book in high school libraries on the Prairies. It was assisted in this project by donations from the Saskatchewan Department of Agriculture and Food; Alberta Community Development; Manitoba Culture, Heritage and Recreation; Saskatchewan Municipal Government, Heritage Branch; IPSCO Inc.; Prairie Farm Rehabilitation Administration (Agriculture Canada); and the Manitoba Department of Agriculture. Companies and individuals who purchased a special edition of the book as part of the fund-raising for school distribution were Gavin Semple, Brandt Industries, Ltd., Regina; Del-Air Systems Ltd., Humboldt; John Buhler Inc., Winnipeg; Leon and Ray Malinowski, Leon-Ram Enterprises Inc., Yorkton; MacDon Industries Ltd., Winnipeg; and Ross Giles, Regina.

Elise Corbet carried out most of the research, conducted most of the interviews, and shared her knowledge during the writing of the text. Donald Wetherell co-ordinated the project and wrote the text. Despite the assistance of so many, any faults that the book may contain, and the conclusions of the study, are ours alone.

Introduction

Machinery has been at the heart of prairie agriculture. It shaped the agricultural practices of the region and, in many respects, created the opportunity for rapid European settlement in the late 1800s. Technological change and innovation have been central in this process and often were justified by the aridity of the region, the short growing season, and shortages of labour. These problems were addressed through different technology, by changing field cultivation practices such as the use of summerfallow and, later, of trash-cover farming, and by progressive developments in plant breeding.

In the beginning, the machinery that so profoundly shaped prairie agriculture was usually designed and manufactured outside of the region. Farmers not only often found it expensive and inferior in quality, but of greater importance, its use also sometimes brought disastrous consequences for the soil.[1] In Canada, this problem was unique to the Prairies, and its solution stimulated an extraordinary level of thought and experimentation. Some of the technological solutions developed in this process of adapting to the land and climate were applied by regional manufacturers, especially in tillage and seeding equipment, which became the backbone of the prairie farm equipment industry. In this sense, the contemporary farm equipment industry on the Prairies was born in the dust bowl of the thirties.

While the challenges presented by conditions during the depression years were important, the industry grew from over a century of innovation and effort in the region. The Canadian prairies are often seen only as producers of acres of wheat, and, except for Alberta, as possessing an economy little changed from before World War II. Yet the manufacturing of farm equipment became a significant part of the prairie economy after the war. The limited fortunes of farm implement manufacturers before World War II were not, however, due to a lack of imagination, invention, or effort. The question is then obvious: If prewar manufacturers found little success in contrast to postwar manufacturers, what changed?

The customary view of prairie economic history holds that diversification of the economy was held back by national economic and transportation policies. But an examination of the farm equipment industry reveals that regional economic development in this sector was more complex than this view suggests. Clearly, the postwar success was part of a series of interconnected events in an evolving regional and national economy. Changes in the local market, in the use of farm equipment, in the dominant players in the industry, in methods of fabrication and industrial infrastructure, as well as in government policy, shaped the fortunes of the prairie industry. Manufacturing is a process—a range of interconnected activity and circumstances—dependent on infrastructure such as transportation networks, energy, and supply of materials. It also depends on a ready market and individuals with varied skills to fabricate, manage, and market the products.

This study broadly discusses significant prairie contributions to farm equipment design from the late nineteenth century until about 1980. The application of this technical innovation in prairie manufacturing has been of particular concern. The study explores the historical evolution of implement manufacturing in the region to isolate some of the guiding forces in its history and provide the historical context for its development. In this way, it explores one aspect of the history of prairie economic development.

The study is divided into two parts. The first part provides a broad outline of the history of prairie agriculture, with particular reference to the forces that had an impact on the purchase and use of machinery. It also looks broadly at the players who manufactured and sold such machinery in the region, as well as at related issues such as farmer and government response to the industry and the prairie industry's evolution and maturation as an economic force in the region. Prairie manufacturing of farm equipment was characterized by a number of small firms filling highly specialized niches left unserviced by larger firms operating from outside the region. These large companies, called full line or main line firms, set the standards for the whole farm equipment sector and defined the marketing and distribution approaches followed in the industry.[2] Farmers' response to the full line companies was complex and resulted in various initiatives, ranging from government regulation of the industry to the creation of farmers' co-operatives. Smaller manufacturers were active on the Prairies from the late nineteenth century, and while their financial and corporate success was limited before World War II, their evolution reveals a great deal about the history of economic diversification and manufacturing on the Prairies before World War II. Although they were a significant precedent for the companies that followed, the later development of the prairie industry reflected changed circumstances. The industry showed not only increased production and expanding sales, but it also demonstrated a growing maturity that recognized the interdependence of its members, leading to the formation of the Prairie Implement Manufacturers Association in 1970.

The second part of the study examines important technical innovations in equipment design on the Prairies and places these inventions in the context of the agricultural practices in which they were used. Many of these designs had an impact on regional manufacturing of machinery. We begin with a discussion of the transition to steam and internal combustion engines, which were so central in the evolution of prairie agriculture and rural society. A conventional categorization of farm implements which divides them into tillage, seeding, harvesting, and haying equipment has been used. We have broadened the haying equipment category to include other equipment related to the feeding and care of livestock and poultry. The study concludes with a discussion of possible future trends.

—PART ONE—

The Prairie Region

Agricultural implement manufacturing on the Prairies developed within the region's distinctive climatic and geographical conditions, as well as its political, economic, and agricultural evolution. The prairie region, or the western interior of Canada, consists of the present provinces of Manitoba, Saskatchewan, and Alberta, totalling almost 2 million square kilometres. This huge and varied area contains well-defined variations in soils, precipitation, and mean temperature. Large portions are not arable, but, generally, the Prairies are characterized by bands of brown, black, and grey soils, relatively short summers, cold winters, and low precipitation. Essentially landlocked, the region is separated physically from central Canada and the lower British Columbia mainland by the rugged and sparsely populated Canadian Shield to the east and the mountains to the west. Other than limited ocean outlets through Hudson Bay or to the Arctic through the Mackenzie River, movement within Canadian territory to or from the region must deal with these geographical conditions.[1] Within these natural conditions, economic and historical forces have directed development towards wheat monoculture, carried out by individual farmers working their own land.

By Royal charter, the Hudson's Bay Company owned all lands draining into Hudson Bay, and this included almost all of the prairie region. Its earliest efforts were confined to posts on the shores of Hudson Bay to which furs were brought from the interior. The settlement of the prairie region by Europeans began in the late eighteenth century with the establishment, in response to Canadian trade competition, of the first inland post by the Hudson's Bay Company at Cumberland House in 1774. Given the nature of the trade, activity was concentrated in the northern wooded areas. By the early nineteenth century, however, the plains and park-belt areas were helping supply food for the fur trade. Under the patronage of Lord Selkirk, the first Scottish settlers arrived in 1812 at Red River, later the site of Winnipeg. While also serving as an economic focus of the fur trade, this was the earliest strictly agricultural effort by Europeans in the region, and the settlement soon became home to Métis settlers and retired Hudson's Bay Company employees, as well as the original Selkirk settlers.

The Selkirk settlers' agricultural efforts are often credited as the first in the region. However, the Native peoples in the western interior had carried on agriculture long before the arrival of Europeans. Although little is known about Native agricultural practices, some tribes were producing corn and other cereal crops, as well as tobacco.[2] It is likely that the Native peoples found, as Europeans would later, that the land was hard to farm. No doubt, they too lost crops to drought, hail, early frosts, and the other natural calamities that have become part of the prairie agrarian tradition. Certainly the Selkirk settlers faced these natural difficulties, as well as the hostility of the Nor'West traders who saw them as allies of their enemy, the Hudson's Bay Company. Agriculture at Red River was also held back by a lack of export markets, while the buffalo hunt and the fur trade drew attention away from agriculture. So uneven was the success of agriculture at Red River that some observers contended that agriculture would forever be a marginal activity in the region.[3]

Opinion about the agricultural potential of the Prairies began to change under the influence of European exploration of the region and Canadian pressure for its annexation. The expedition of Captain John Palliser in 1857 concluded that large portions of the western interior were suitable for agriculture, although one area, later known as Palliser's Triangle, was characterized as a desert. While this finding was used in support of the commonly held belief that the area was unsuitable for agriculture, by the last quarter of the nineteenth century these beliefs were coming under mounting attack. For example, the writings of John Macoun, called a "propagandist" by one historian, helped dispel the view that a portion of the western interior was a near desert.[4] As events in the twentieth century were to prove, some of this was little more than the harmful propaganda and wishful thinking of Canadian expansionists, but it also contained, as later events would also prove, enough truth to be justifiable. Yet, at the heart of the debate were the expansionist dreams that were in the air in Canada. The confederation of three British North American colonies into a single country in 1867 was partly due to the demand by central Canadians for expanded markets and territory. This ambition included the hope that the West, including the prairie region, would be annexed and integrated into the Canadian economy and saved from American expansionism. This was achieved in 1869 when Canada purchased the Hudson's Bay Company lands, including almost the whole of the present Prairie Provinces.

Canada worked quickly to integrate the new territories. In the first few years, resistance by the Métis in Red River to the Canadian take-over was quashed; a postage-stamp-sized province, Manitoba, was created; and a government was established for the rest of the Prairies, now known as the North-West Territories. By these actions, Canadian laws were extended to the region. Between 1871 and 1877, treaties for most of the land south of the North Saskatchewan River and between the Rockies and the Red River were also signed with the Native peoples. The greatest project, at least in contemporary eyes, was to link this western interior with the rest of Canada. Beginning in 1873, the Canadian government began exploring the possibility of a rail line to provide an economical link between east and west, to stimulate settlement of the West, and to provide a mechanism for moving prairie products to world markets and, in turn, Canadian manufactured goods to the Prairies.[5] While political scandals, change in government, and economic depression conspired against the project, by 1880, it had been launched with the formation of the Canadian Pacific Railway Company (CPR). By 1883, Winnipeg, Regina, and Calgary were linked by rail; by 1885, the line north of the Great Lakes was completed; and by 1886, rail had reached Vancouver.

In addition to these developments, a land system was implemented in 1872 to guide subsequent European settlement. Like the CPR, land policies were designed to encourage settlement. Canada used a survey system that methodically surveyed land into numbered units. A simple and efficient system, it anticipated rapid settlement and also provided a precise method of distributing lands to the Hudson's Bay Company and to the CPR. Each received large amounts of public land, as part of the purchase agreement in the case of the Hudson's Bay Company and as a subsidy to construct the line in the case of the CPR. Additional lands were set aside for future sale to support education, and others were distributed to homesteaders for a $10 filing fee. In this way, prairie farmers could acquire inexpensive farm land and could also purchase additional land from the CPR, the Hudson's Bay Company, and, in some cases, the Crown. And, after 1908 a homesteader could "make an interim claim upon an adjoining quarter section" and purchase it for $1 per acre, a policy that encouraged larger farms. This trend towards larger farms, which mechanization and economies of scale also increasingly encouraged, was an early feature of prairie agricultural life.[6]

Canadian transportation and land policies served to confirm the prairie region's role as an economic hinterland of central Canada. The region's role in the national economy was to produce primary products like wheat and to consume manufactured products produced elsewhere. Tariff policies were another aspect of national development that had an impact on the Prairies. Before 1879, tariffs on goods imported into Canada had been imposed in order to produce government revenue. In 1879, however, the objective of tariff policy was changed by the government of John A. Macdonald. Tariffs were now set to provide Canadian manufacturers with protection from foreign competition, which, in turn, would stimulate manufacturing in Canada and, equally important, encourage interprovincial trade and the growth of the east-west economy. This would then increase east-west railway traffic and make the railway more profitable.[7]

Even though the infrastructure for European settlement—government, law and order, land policies, tariffs, and the railway—were in place by the mid 1880s, settlement was slow to take off. By 1891, the population of the entire North-West Territories was just over 254,000, more than half of which was in Manitoba. The whole of Canada attracted only 1.5 million immigrants in the entire period from 1867 to 1899.[8] On the Prairies, most early settlement was in Manitoba, which became a testing ground for prairie agriculture and settlement. By 1870, settlement in Manitoba followed the river fronts along the Red and Assiniboine rivers, as far south as Pembina on the Red River, and to Portage la Prairie on the Assiniboine. In 1871, Winnipeg had a population of between 100 and 200, but by 1886, it had increased dramatically to about 20,000. Outside the urban centres, the most significant change was that settlement now no longer clung only to the rivers but had moved onto the open plains, although most settlers still tended to select sites in wooded areas, with higher moisture and brown soils, when possible. However, Mennonite settlers in the late 1870s had demonstrated that the open plains of Manitoba could be farmed successfully, and their example was further reinforced by a growing shortage of land in wooded areas.[9] So although Manitoba's agricultural settlement continued to be concentrated along the river fronts, a new pattern was becoming apparent by 1886. Settlement clustered in four main areas: southwest of Winnipeg around the present Manitou and Morden; in Brandon and district; south of Brandon around Souris and Boissevain; and north of Brandon around Minnedosa and Neepawa. Yet of the 4 million claimed acres of land in Manitoba, only about 18 percent was cultivated.[10]

Titled "A characteristic western farmyard," this 1890 photograph reveals the small amount of equipment used in early prairie agriculture. (Ernest Brown Collection, B720, PAA)

While settlement gradually spread westward beyond Manitoba's borders, the coming together of various historical forces in the late nineteenth century stimulated a dramatic upswing in the number of immigrants and European settlement of the western interior. The filling up of the American west turned settlers' attention to the Canadian Prairies, which truly became the "last best west." By the turn of the century, settlements had been firmly implanted in the south-eastern part of the future province of Saskatchewan, which, while drier than the park-belt areas, still had adequate moisture. Settlement had also occurred in southern and central Alberta, and while large sections of the region were as yet unsettled by Europeans, the settled areas were tied together by railways. By the late nineteenth century, it was clear that agriculture on the Prairies had to deal with distinctive challenges: in the south, it was aridity and a short growing season, and while the more northerly areas enjoyed slightly more precipitation, they had an even shorter growing season than in the south.[11] The decade between 1901 and 1911, however, was a period of higher than normal rainfall on the Prairies and high world wheat prices. This plus the economic growth associated with settlement and railway construction created a climate of optimism and prosperity. In this decade alone, the region's population increased from about 419,000 to about 1,328,000. Confidence in the future was so great that in 1908 the federal government opened the much drier areas in south-western Saskatchewan and south-eastern Alberta, the so-called Palliser's Triangle, for settlement, a policy that would prove to be disastrous for many of those who settled there.[12]

Between 1908 and 1918, prairie agriculture flourished, stimulated by the worldwide economic expansion before World War I and the war years, which brought higher prices for wheat and other produce. Widespread adoption of earlier maturing wheat encouraged the wheat economy even more. Yet, by 1917, another aspect of the future had become apparent as well: wrong agricultural practices and

By 1920, as this photo of a farm near North Battleford, Saskatchewan, shows, farms had become better equipped and more prosperous than in the settlement period. (Harry Pollard Collection, P697, PAA)

technology, combined with drought, led to the start of farm abandonment in the Palliser's Triangle. The years after the war were dry ones, and productivity declined and prices fell. By the mid 1920s, however, conditions improved, and immigration, which had been low, picked up. New areas, such as the Peace River Country in northern Alberta and northern fringe areas, such as those around Nipawin, Saskatchewan, drew increased settlers. Acreage under cultivation on the Prairies continued to grow. The 6 million acres of cultivated land in 1901 had expanded to 34 million acres by 1916 and, by 1929, had reached 60 million acres. It was an economy dominated by grain, especially wheat, although the Depression of the 1930s would present a profound challenge to the wheat economy and its organization.[13]

The Prairie Agricultural Economy

In the late nineteenth century, there was a great deal of debate about the best type of economic development for the Prairies. Most assumed that agriculture was the key to both the region's prosperity and national economic priorities. But what sort of agriculture? From the beginning, mixed farming, in which production of field crops was combined with livestock and poultry, had many supporters. It was believed to be more stable than monoculture and to bring the greatest number of economic spin-offs at local, regional, and national levels. Between 1870 and 1918, mixed farming had powerful and tireless advocates, including the territorial and then provincial governments, town businessmen and promoters, agricultural theorists, and the farm press.

Initially, these expectations were confirmed. In Manitoba, agriculture in the late nineteenth century tended to be dominated by mixed farming. But, although mixed farming subsequently developed elsewhere on the Prairies, especially in Alberta, the overall regional trend was to wheat farming. In 1910, wheat acreage was largely found in the Red and Assiniboine River valleys and along the CPR mainline as far west as Regina. This totalled about 3 million acres, and acreage in wheat subsequently increased. While Manitoba and Alberta were never dominated by wheat growing like Saskatchewan, farmers in all provinces responded less to theories about idealized agriculture and more to practices that meant more profits and a better lifestyle. And grain farming seemed to fit those ambitions best. During World War I, this pattern of prairie agriculture was firmly established. There was significant expansion of acreage under cultivation during the war, both in response to high prices and to government programs that encouraged it. Most importantly, because of high demand for wheat during this period, "the fatal prairie habit of dependence on a single export commodity" was established more firmly.[14] By 1926, there were 22 million acres in wheat in the three Prairie Provinces.[15]

Although the World War I period was crucial in establishing wheat as the primary crop on the Prairies, it had long been a favoured crop because of high European demand. With European industrialization and urbanization in the nineteenth century, and the accompanying decline in production of cereal grains, greater and greater amounts had to be imported. Of these, wheat was most important. Public demand was high for flour made from hard spring wheat because it produced very white bread. This was not a mere consumer preference; it was a product of centuries of European views that white bread, so uncommon a product, was a sign of wealth and status. The food of the aristocracy and of story-book princesses could now be realistically demanded by all. One of the greatest sources of this hard spring wheat was North America. This demand created an expanded market for Canadian wheat, which stimulated prairie agricultural development. The prairie agricultural economy was tied to Europe now, and as any economy dependent on foreign demand, it experienced periods of rapid growth and sudden contraction in step with European events.[16]

High demand during World War I demonstrated one side of this pattern of dependence, and by 1918, wheat was "Canada's most important single export in dollar value."[17] This condition lasted until after the war, and by the late 1920s, 60 percent of the prairie farmer's cash from field crops came from wheat. By 1930, wheat production in Canada totalled 500 million bushels (most of it produced on the Prairies), and the nation's wheat exports commanded 40 percent of the world's wheat export markets.[18] Yet, in the late 1920s, rising production and high domestic support prices in Italy, Germany, France, and other European countries (the United Kingdom excepted) combined with quotas to affect Canadian sales negatively. In an irony apparent only with hindsight, the peak in prairie wheat production came in 1928 on the eve of the Great Depression. In 1929, the interconnectedness of the prairie and world economies was further shown by the dramatic collapse of international agricultural prices and markets.

In 1932, the price of wheat dropped to extremely low levels. Though the international economy began to recover in 1933, the drought hung on in the Prairies. This, together with continued low commodity prices, made the farm crisis desperate. Production fell; the average prairie wheat crop of 350 million bushels per year in the 1920s decreased to an average of 230 million bushels per year from 1933 to 1937. Farm incomes, which had been about $500 million per year in the late 1920s, had dropped by more than half by the early 1930s.[19] Farm abandonment was one result, and in 1936, there were 14,000 abandoned farms on the Prairies, mostly in southern Saskatchewan and Alberta. Some people moved to new homes in northern Saskatchewan and Alberta, but large numbers left the region for good. During the 1930s, 250,000 people migrated from the Prairies, while about 50,000 moved to new homes further north. Not only were the human costs high, so too were the capital losses that the upheaval represented.[20]

The Depression also brought about an increased emphasis on crops other than wheat. Mixed farming had always been more significant in Manitoba than in Saskatchewan, and by 1921, wheat occupied less than 50 percent of Manitoba's field crop acreage. Partly because of soil depletion and poor agricultural practices, Manitoba farmers had begun shifting away from wheat in the 1920s, a pattern the Depression reinforced.[21] Wheat also began to give way to barley and forage crops in central Alberta in the 1930s. But the shift away from wheat was neither smooth nor continuous. With the outbreak of World War II in 1939, farmers responded to anticipated higher demand for wheat by increasing their wheat acreage. The romance between prairie farmers and wheat, their first love, seemed on the verge of renewal. German military conquests resulted in the complete closing of European export markets in 1940, however, and markets for Canadian wheat contracted. The Canadian government responded by attempting to get farmers to reduce acreage in wheat in favour of more diversified production, and this policy continued until the end of the war. Although bumper crops in 1940 and 1942 meant increased wheat production despite a decline in acreage, by the end of the war, mixed farming had gained ground, even though wheat remained an important crop almost everywhere in the region. But as the Saskatchewan economist Vernon Fowke remarked in 1945, these changes, like others in the prairie economy, were "probably neither as fundamental nor as permanent as is commonly thought." As he noted, the change from wheat to mixed farming was "substantially a producers' response to changing cost-price relationships, and will persist or revert depending on what happens to those price relationships in the future." [22]

Events showed that Fowke had correctly pinpointed the dynamic of farmers' cropping decisions. In the next two decades, wheat acreage increased or declined in response to farmers' beliefs about what would be profitable. Nevertheless, by 1965–66, farm diversification had become more characteristic of all provinces except Saskatchewan. There, 65 percent of total farm sales, versus only 25 percent and 31 percent in Alberta and Manitoba, respectively, came from wheat.[23] Alberta had a relatively diversified agricultural economy—grasses in the Peace River area; cattle, sheep, and sugar beets in the southern areas; and mixed farming in central Alberta. Southern Manitoba also had considerable diversity, including extensive production of honey, sugar beets, potatoes, sunflowers, corn, and canning crops. Wheat remained the major crop in Saskatchewan, although mixed farming and alternate crops like rapeseed diversified the economy in the north-central parts of the province.[24]

But this diversification only moderated a basic difficulty that had always plagued prairie agriculture. In the 1960s and 1970s, prairie farmers faced the same problems they had for a century: too small a regional market plus competition from farmers closer to the major centres of population in Canada forced them to sell in foreign markets in which they were unable to shape prices. Nor could they control transportation or farm input costs. And, as in the late 1920s, they also faced static or even declining markets, although farm production was increasing.[25]

Farm Expenditure

The vulnerability of prairie farmers to foreign markets and their lack of control over farm costs seemed to be a hard set of problems to overcome. These difficulties were made sharper because of the economic conditions of settlement and of prairie farming. Agriculture demands long-term investment, and while many migrants had initiative and ingenuity, few had much capital. Although farms could be set up initially for a relatively small outlay, ongoing success required capital.[26] A well-equipped farm required an expenditure of up to several thousand dollars, and this capital had either to be brought by the migrants or earned on the frontier. To earn enough cash to keep their farms solvent, many farmers before World War I worked part-time on the railway, in mines, and in logging camps, or as farm hands for wealthier neighbours. This was partly the result of low initial capitalization, but varying commodity prices also created a need for off-farm work. During World War I, when prices were high, the need for this work declined, but depressed prices in the early 1920s re-emphasized the need for additional income. Mechanization during the 1920s also created a demand for capital. Yet the ending of the frontier phase meant less available work on railway construction, and then the 1930s saw decreased government expenditures on roads.[27]

Given these circumstances, it is not surprising that there was a significant failure rate on prairie farms. Although the failure rate in Manitoba has been estimated at only 20 percent from 1870 to 1905, drought and soil depletion between 1919 and 1923 led some to abandon their farms in south-western Manitoba. Similarly, between 1911 and 1931 in Saskatchewan, and between 1905 and 1930 in Alberta, 57 percent and 45 percent, respectively, of farms ended in failure. Personal reasons lay behind some of these failures, others resulted from attempts to farm unsuitable land, and still others were due to the bad luck of beginning to farm on the eve of an economic down-cycle.[28]

In general, these problems revolved around the legacy of too little capital to start up farming, and they were made worse by varying commodity prices and new-found costs. By the interwar years, farm debt was high. In the late 1920s, it was estimated that 30 percent of farm capital was borrowed at fixed interest rates.[29] While some of this debt was due to the purchase of tractors and other mechanization of prairie agriculture, it also increased because of the Depression. The Alberta government estimated in 1932 that "more than three times as many bushels of wheat were required to provide funds for the payment of the same amount of interest as in 1926." Because farmers "had to live," they postponed debt repayments "with the result that the debt burden upon which interest had to be paid grew phenomenally."[30]

The costs of farming before World War II were therefore a blend of various elements. And these farm costs increased further in the postwar years. While many farmers were able to pay off debt because of higher incomes during the war, this was offset to a degree because of greater farm size and increased mechanization. The percentage of farm expenses invested on the Prairies in farm implements and machinery increased from 10.5 percent in 1921 to 16.3 percent in 1941 and 24.4 percent in 1951. By 1961, it stood at about 30 percent. In addition, other input costs such as for fertilizer, chemicals, and seed, all of which were relatively minor before World War II, had also increased. Prairie farms had traditionally required high investment, and after World War II, these capital requirements continued to grow. While there was considerable variation within the region due to varying land productivity and the type of agriculture practised, the average farm investment in Alberta was about $75,000 by the 1960s. The investment for grain farms in the southern Prairies was much higher—upwards of $200,000. These costs were handled through debt and expanded scale of operations because commodity prices had not increased in the same fashion.[31]

Labour was a key component in this changing pattern of farm expenditure. Because it created an important (although not the only) justification for mechanization, it played a central role in the reshaping of prairie agriculture. Before World War II, hired labour was a continual issue on the Prairies. It was difficult to attract workers for the wages that farmers were willing or able to pay for short-term seasonal work. And farmers had a reputation as poor employers. The emphasis in prairie culture on the virtues of land ownership also created a barrier to the development of a stable wage workforce. If it was easy to own land, then most workers wanted to be landowners, and working for others was often looked down upon. As a result, mechanization was stimulated by the belief that it would free farmers from the need for hired labour,[32] and the use of hired labour declined dramatically. Expenditure for hired labour on prairie farms in 1921 was about $75 million, but it had fallen to $45 million in 1946.[33] This extraordinary decrease in what had traditionally been a labour-intensive industry mirrored the massive changes in mechanization that had occurred by the end of World War II.

Despite mechanization, hired labour remained important in prairie farming. These harvest workers heading west took a break somewhere in the Canadian Shield. (Archives Collection, A10499, PAA)

The decreasing role of farm labour as a capital requirement marked an important change in agriculture, especially when it is recognized that the start-up cost of prairie agriculture was not only a matter of farmers' cash input. Their labour, in combination with government land policies, had also provided much of the initial capital for prairie farms. While outside capital poured into the region before World War I for development of railways and other infrastructure, it seems that farms were not the direct recipients of much of this investment. Yet, the homestead system was a form of public assistance in capitalizing the prairie farm. It has been noted that "as productivity of the land was developed, its worth increased and the values so created accrued to the owner. Hence the land grant system provided agriculture with a large part of its original capital input."[34] Similarly, capitalization of farming during the settlement period also benefited indirectly from public assistance because land was used as a resource to finance the economic infrastructure of railways and schools. Additional benefits also came from other direct forms of public assistance for agricultural programs, research and education. The capitalization of prairie agriculture was therefore a mixture of noncash components, such as personal labour on the farm and public subsidies, and cash obtained through the sale of labour off the farm, the sale of commodities, and borrowing.

So although pioneer farming was never as self-sufficient as prairie myth sometimes holds, cash needs in the early part of this century were lower than they were by mid century. As the agricultural economy developed, more and more cash was required. Mechanization, rising costs of land, and increasing size of farms were elements in this greater emphasis on cash in prairie agriculture. It is clear that prairie farmers preferred to substitute capital (in the form of machinery) for labour (in the form of hired workers). And this substitution of capital for labour had a profound impact on the demands that farmers placed on machinery and on the products and services provided by the farm implement industry.

Manufacturing on the Prairies

While the prairie region has traditionally relied on the export of resources, this has—for some observers—always been too limited an economic base. As early as 1883, *The Commercial*, a Winnipeg business paper, argued that "people who only raise and export breadstuffs, and import the bulk of their manufactured goods may without injustice be accused of shiftlessness." In this view, manufacturing, especially the processing of foodstuffs, was a mark of progress. Without it, the prosperity of the region would be "incomplete." Three years later, the newspaper repeated the argument, once again rejecting the assumption that "the Northwest must ever remain a country of agriculturalists with a few storekeepers sprinkled over it."[35]

The following century saw views like this continuing to be endorsed by numerous boards of trade and town and city councils. While there was some increase in manufacturing, especially in food processing, the ambition for an economy that combined agriculture with manufacturing was largely unrealized. Despite considerable population growth and the development of the region's transportation and urban network, by 1900, manufacturing on the Prairies was still a distinctly minor part of the economy. By 1907 in Saskatchewan, it consisted of flour and grist mills, creameries, and plants that processed and fabricated building products.[36] By about 1910, however, the region's economy had broadened somewhat because of the tremendous growth of population and settlement. World War I also brought significant expansion, and by 1920, industrial production for the three provinces was valued at $300 million. Much of this activity continued to be in food processing and building supplies.[37] While the Canadian Manufacturers

Association did not have any members on the Prairies in 1900, it now had 500, a significant measure since only firms employing at least five workers in their mechanical department were accepted for membership. Most of this activity was concentrated in Winnipeg, which had become the major metropolitan centre of the region.[38]

By the mid 1920s, food processing was still the leading prairie industry. Flour milling was the most common industry in Saskatchewan and Alberta and was second only to meat packing in Manitoba.[39] However, there was much optimism that manufacturing was about to take off. In 1927, it was observed that the value of industrial production had increased substantially in the previous two years and a further increase was likely. National and international interest in western resources was growing, and the supply of electricity was increasing. These factors gave "confidence that in a few years more the industrial activities of the Prairies will be a factor of importance in the economic life of the whole Dominion."[40] This forecast seemed confirmed in 1928 when General Motors opened an assembly plant in Regina.[41]

The General Motors plant, however, enjoyed limited success. It operated only intermittently until 1939, and the problems created by the Depression were a factor in its failure. Similarly, the economic collapse of the 1930s set back diversification of the prairie economy in general. Manufacturers supplying the prairie economy in the 1930s were in a precarious position indeed—the impact of the Depression was not restricted to farmers but extended to all who depended upon them. In urban areas, light industry as well as the wholesale and service sectors declined in the 1930s.[42] The war years that followed restricted new development to war-related activities. While the immediate postwar years saw increased manufacturing activity, it remained a small part of the prairie economy and tended to be restricted to food processing and light manufacturing, usually related to the regional market. In 1951, for example, manufacturing in Manitoba was still dominated by the food and beverage industry which accounted for 45 percent of the province's manufacturing output. This manufacturing sector became less important, however, and by 1966, it stood at about 33 percent of total output. One industry that had grown during this period was farm equipment manufacturing. While this indicates that manufacturing in Manitoba had undergone change in the two decades after World War II, it remained largely tied to primary industry.

The nature of this primary industry, however, was changing. The agricultural dominance of the prairie economy slipped after World War II with growing urbanization and major shifts in resource extraction towards oil, potash, and other minerals. This transition is a mark of the postwar years, but the region's fundamental dependence on primary resources remained. Despite high growth in the 1960s, especially in Alberta, an economy dominated by production and extraction of raw materials meant that the traditional prairie economic problems endured. Prairie manufacturing was oriented to regional markets and was dependent on primary industry. Manufacturers experienced declining sales because of poor markets for the primary industry they served, whether it was agriculture, oil and gas, or mining.

The region's traditional economic quandary persisted. Diversification had lessened the traditional dependence on agriculture, but the regional economy continued to be dependent on primary resource production and processing, whose sensitivity to international demand created unstable employment. Most importantly, consumption and final manufacturing of prairie resources were largely still carried on elsewhere, and most basic commodities were exported in raw form, creating economic activity elsewhere.[43]

By the 1970s, as in the preceding century, manufacturing in Canada was concentrated in the Quebec-Windsor corridor. Its head start in manufacturing gave it national strength, and the concentration of wealth and population in central Canada gave it additional economic and political power. On the Prairies, manufacturing continued to be typified by light industry and custom manufacturing for industry involved in resource extraction. This was reflected in prairie employment patterns. In 1961, 10 percent of Ontario's labour force and 14 percent nationally were employed in primary industry, that is, the production of raw materials, versus 30 percent in the Prairies. This did not, however, translate evenly throughout the Prairie Provinces. In Manitoba, "the value added by manufacturing" was greater than "that contributed by primary industry," while Saskatchewan continued to rely on agriculture, and Alberta on petroleum. At the end of World War II, all Prairie Provinces had depended on agriculture for the largest percentage of provincial production, but by 1970, only Saskatchewan, with the most narrow-based manufacturing sector in western Canada, still did so.[44]

The reasons for the limited role of manufacturing in the prairie economy have been commented on and debated since the late nineteenth century. In the 1880s, the reason most commonly cited for the lack of development in manufacturing was that land speculation (which was periodically rampant in the region) drove up the cost of living and wage rates.[45] In 1914, much the same argument was made with the contention that high wages, caused by expensive housing resulting from land speculation, increased labour costs. Additionally, it was claimed that the protective tariff increased the cost of living and therefore stymied industrial development.[46] While these were typical arguments throughout the interwar years, it became more common to argue that the lack of an extensive labour supply and investment capital, plus competition from well-established factories elsewhere, held back the development of prairie manufacturing.[47]

After World War II, these explanations were refined by an increased recognition that prairie economic development was the product of a number of natural and political factors. This view was typified by the Saskatchewan government in 1976 when it argued that the province's manufacturing was relatively undeveloped because of interconnected constraints—the province was landlocked, it was distant from major North American markets, and it had a small population. National economic policies had also worked to concentrate manufacturing in Ontario and Quebec since Canadian freight rate and transportation policies had discriminated against the growth of resource processing near the source of supply.[48]

The view that limited manufacturing on the Prairies characterized the region's unequal relationship with the national economy has been especially widespread since World War II. Historians have tended to support the argument that regional manufacturing was restricted by the operation of metropolitan forces and by federal economic policies that favoured central Canada. Terence Fay has endorsed this view in his study of farm implement firms in Winnipeg. He argues that farm equipment manufacturing did not develop in Winnipeg between 1876 and 1926 because of the working of the Canadian tariff and freight rate policies, the concentration of economic and political power in central Canada, and the efficiency of outside industries. These forces, he contends, "inhibited the development of an indigenous machine industry." [49]

The tariff is commonly said to have inhibited prairie manufacturing by force-feeding central Canadian manufacturers and enabling them to entrench their head start in manufacturing. As a corollary, the costs of this artificial growth were borne by all Canadian consumers. A related argument is that Canadian freight rate policies discriminated against the development of manufacturing in the West because it was cheaper to ship manufactured goods to the Prairies from central Canada than it was to ship the same items eastward. As Gerald Friesen has noted, critics of Canadian freight rate policies contended that since the railway was built for national purposes, the whole country should have helped to pay for it, and that "the fundamental principle of Canadian transportation policy should not be regional discrimination but rather should be national equalization of railway rates." [50] Nonetheless, cheaper rates for manufactured goods shipped westward were in effect, and farm implements serve as one example among many. The Crow's Nest Pass Agreement of 1897 provided lower freight rates for iron, wire, "settlers' effects," and farm equipment of all kinds moving west. In fact, lower rates than these were in effect from 1897 to 1922. When rates began to rise in the early 1920s, however, the CPR refused to honour the rates set by the Crow's Nest Pass Agreement and succeeded in having its terms renegotiated.[51] In the meantime, however, a rate higher than the original Crow rate but lower than what the CPR wanted was set for farm machinery in 1919. In 1939, an increase in this and other rates created much indignation both among prairie farmers and the provincial governments. As a result, the lower freight rates on farm equipment "in contrast to other commodities" were reinstated. In 1950, this lower rate for farm machinery was abolished.[52] There were no lower rates for shipment of farm equipment manufactured on the Prairies to eastern markets.

A negative relationship between freight rates and the development of prairie manufacturing became a standard part of prairie political rhetoric by the 1970s. In 1973, for example, the prairie premiers blamed the Canadian freight rate structure for holding back the diversification of the region's economy by restraining the development of manufacturing. Critics of this view have in turn argued that, given the basic characteristics of the railway industry and competition among transportation sectors (such as between rail and truck), rates have not been discriminatory. This argument has concluded that the role of freight rates in hindering the economic diversification of the Prairies "is very small, if it exists at all." [53] While freight rates and tariffs, and their role in the economic diversification of the prairie economy, remain subject to debate, some observers of the region contend that neither tariff nor transportation policies can be blamed wholly for the lack of manufacturing on the Prairies or for its continued reliance on resource production. Arguing that prairie economic problems are in part due to low population and distance from heavily populated and wealthy markets, the economist Kenneth Norrie has suggested that these problems were the natural and inevitable results of a sparsely populated region's fate within a market economy.[54]

These general arguments about prairie economic development provide one framework for understanding the evolution of farm equipment manufacturing on the Prairies. Many observers of the prairie economy and its history have tended to see the region as existing in the grip of outside forces, especially national economic policies. Yet while local governments on the Prairies avidly promoted industrial development of all sorts, the provincial governments, with the partial exception of Manitoba, historically stressed development of farming over industrial growth. Only after World War II did they consistently begin to promote a more diversified economy.

For farm equipment manufacturing, the economic and technical infrastructure to support such an industry was limited before World War II. Fabrication techniques used in heavy manufacturing also created a barrier to prairie manufacturing. As well, the North American industry was dominated by large corporations based in Ontario and the United States, and they set the standard for all farm equipment manufacturers. Only at the beginning of World War II did changing prairie agricultural practices stimulate a need for equipment specific to prairie conditions. Given the small market size and the different concerns of the dominant manufacturers of farm equipment, this unique equipment had to be invented and manufactured in the region. The emergence of a prairie farm equipment industry was therefore a response not only to economic forces but also to fundamental change in agriculture and the region's industrial infrastructure.

The Farm Implement Industry

Manufacturing of farm equipment on the Prairies developed within the context of an industry dominated by a handful of large corporations outside the region. Since they produced a full range of equipment, they were known as full line companies, producing equipment from the most complex sort such as tractors to the simplest such as harrows. In contrast, as discussed in a later chapter, smaller specialized firms producing only a few lines of equipment were called short line companies. The full line industry was characterized by integrated manufacturing, distribution, and sales. In Canada, its growth was partly stimulated by tariffs that made foreign goods more expensive. After free trade in farm machinery was established between Canada and the United States in 1944, the industry increasingly operated on a continental basis.

Agricultural equipment represented a large annual expenditure in Canada. The substitution of capital for labour through farm mechanization was beginning to characterize prairie agriculture before World War I and during the interwar years, but it increased greatly during the 1940s. Canadian farmers as a whole spent $30 million for farm equipment in 1939, $82 million in 1945, and $269 million by 1953. Consistently, about two-thirds of such expenditures was on the Prairies, making it the largest market in Canada for farm equipment. The greatest increase in the cost of equipment also came in the late 1940s; between 1946 and 1951, it increased 58 percent. The value of farm equipment increasingly represented a greater percentage of total farm capital. In Saskatchewan, for example, it increased from about 7 percent in 1911 to just over 18 percent in 1946, and increased further subsequently.[1]

While the size of the Canadian market was considerable, it was dwarfed by that of the United States. Between 1939 and 1953, for example, sales there ranged between 6 and 11 times Canadian levels.[2] Both markets, however, were highly variable. Farmers' expenditure on farm equipment was related directly to their income. And since farm income historically has had greater variation than income generally, the farm equipment industry faced chronically unstable demand. Yet, even though it was vital to agriculture and received extraordinary attention from farmers and politicians, the industry was a small sector of the Canadian economy. In the mid 1950s, it represented only about 1 percent of Canadian industrial output and ranked 30th out of 39 manufacturing sectors in Canada.[3]

The Full Line Companies

The large full line companies in both Canada and the United States grew through the consolidation and merger of smaller firms, often short line companies that had produced specialized lines of equipment.[4] Before about 1900, most farm equipment manufacturers in the U.S. were short line companies, but full line firms began to emerge in the early years of the century. This trend was encouraged because short line firms found it both difficult and expensive to distribute their products due to the seasonal and cyclical nature of the farm equipment market. There was also a tendency to monopoly, resulting partly from scale of production as well as from conscious efforts by manufacturers to overcome competition. This was the case with the first full line company in the United States, International Harvester Company (IHC). It was created in 1902 by the amalgamation of the five largest American manufacturers of harvesting equipment. The goal was to overcome competition and increase profits—an objective all but ensured since the new company manufactured 90 percent of grain binders and 80 percent of mowers produced in the U.S. While it initially specialized in harvesting equipment and was therefore still a short line company, although a large one, by 1911 it had increased its production lines and had become a full line company.[5]

IHC set a trend that came to characterize the full line industry in the United States. There, seven full line firms dominated the farm equipment industry by 1936, with the result that a few companies were able to "determine the conditions under which all manufacturers and dealers must operate." As the U.S. Federal Trade Commission concluded in 1938, "similarity of policies respecting prices, terms and competitive practices, to some extent tends to result whether the policies are determined by the companies individually and competitively or by understandings or agreements." The "price leaders" of this system were John Deere and Company and IHC. Other manufacturers delayed announcing prices until Deere and IHC had published their prices. This, as well as the exchange of price lists among individual manufacturers, contributed to price rigidity. Standardization of design also contributed to uniform pricing "because implements having no distinguishing characteristics cannot well command higher than ordinary prices except through established custom or special sales promotion." By the 1930s, these conditions and practices were "substituted for the older, cruder and more obviously illegal agreements" by which manufacturers had controlled prices. Even in the Depression of the 1930s, while sales fell drastically, the industry was still able to maintain its prices overall.[6]

In most respects, the nature of the farm implement industry in Canada was the same as in the United States, although market dominance by a few companies occurred slightly earlier in Canada. Before the 1860s, farm implements in Canada were supplied almost entirely by small Canadian firms relying on imported technology, primarily from the U.S. The increased use of machinery in farming along with expanding agricultural markets in the 1870s led to the expansion of the Canadian industry. This was especially the case after 1875 when the tariff on imported farm machinery was increased substantially. It rose from 10 to 17.5

percent in 1874, and by 1883, it was 35 percent, where it remained until 1894 when it declined to 20 percent. It stayed at this level until 1907, when further declines began, and by 1930, Canada's tariff stood at between 6 and 7.5 percent on most items.[7]

These tariff levels were of great significance to the Canadian industry. Occurring at a time of major expansion in Canadian agriculture because of the opening of the West, tariff protection provided important assistance to Canadian manufacturers, especially in central and eastern Canadian markets. And perhaps not coincidentally, Canadian companies after 1885 began to merge into full line companies. The most important merger was between A. Harris, Sons and Co. and the Massey Manufacturing Co. in 1891 to form Massey-Harris. This merger gave the new firm control of about one-half of Canadian production of agricultural equipment, and subsequent acquisitions further expanded its domination. The other Canadian full line manufacturer that emerged before World War I was the Cockshutt Plow Company which, with its 1909 purchase of Frost and Wood, added harvesting machinery to its lines. Like Massey-Harris, additional acquisitions broadened its lines. The other full line company on the Canadian scene was IHC, which established a branch plant in Canada in 1903 in order to jump the tariff barrier. This plant also manufactured for export since Canada enjoyed a preferential tariff rate in South Africa and, on some lines, in Australia.[8] By 1909, a triumvirate made up of two Canadian and one American corporations ruled the Canadian farm equipment industry. In addition, John Deere set up a manufacturing plant in Canada in 1911, but it was a comparatively small operation.[9]

Neither the Canadian nor American industry marketed products for its domestic market alone. From 1881 until 1929, exports of farm equipment from Canada increased from less than 10 percent of the value of production to almost 50 percent.[10] Most of these exports went to South America, the British Empire, and Europe, and these markets remained the destination of most Canadian farm equipment exports until World War II. Up to 1913, the year most import duties on farm equipment were abolished, the United States was a highly protected market. Until 1897, the American tariff on imported farm equipment was 45 percent, a rate never matched in Canada, and Canadian exports to the U.S. were understandably low. As a result of the American tariff barrier, Massey-Harris set up a plant in the U.S. in 1910, doing there what American firms like IHC were doing in Canada.

Despite the tariff, imports of farm equipment from the United States into Canada rose rapidly after 1901. This reflected the opening of the Canadian prairies where American manufacturers, mainly located in the Midwest, enjoyed a freight advantage over central Canadian firms. Even when Canadian firms gained a freight cost advantage to prairie points in 1919 with adjustments to freight rates, imports of American equipment to the Prairies did not decline. Indeed, they were increasing rapidly because tractors, then a new machine, were being imported from the U.S.

In an effort to increase wartime food production, the duty on tractors valued at less than $1,400 was dropped in 1918. This effectively abandoned the Canadian market to American firms, and few Canadian efforts at tractor manufacturing were successful. Reflecting this continental approach to tractor manufacturing, Massey-Harris purchased a tractor plant in the United States in 1927, from which it supplied its Canadian market.[11] The manufacture of tractors in Canada and their export to the U.S. was not realistic at this time because tractors required expensive production facilities, making large production runs essential. The Canadian market was appreciably smaller than the American one, and a solid export market would have been necessary for the manufacture of such expensive equipment in Canada to be profitable. And there were reasons why the U.S. was an unreliable export market. As T.A. Russell, president of Massey-Harris, asserted in 1936 in his testimony to a House of Commons inquiry into farm equipment, American political history demonstrated the instability of the U.S. as an export market. He asserted that "if anything began to seriously interfere with domestic business in the United States," the American government traditionally had acted quickly to "curb such competition." Moreover, Massey had "found a definite prejudice against Canadian-made goods in the United States," although Canadians did not mirror this attitude to American goods.[12]

Indeed, the Canadian tendency to favour American farm equipment was particularly pronounced on the Prairies. The large number of American settlers in the region perhaps partially explained this attitude, but United States equipment was also better suited to the Prairies. In 1893, it was argued that American firms produced equipment suited to "their own prairie country similar in requirement to our own." [13] This challenge was hard to meet, and in 1902, it was remarked that Ontario firms were disadvantaged on the Prairies because they "do not study the requirement of the market they attempt to cater to. They try and persuade the westerners that they need machinery such as is used in Ontario. But the westerner knows they do not; they know just what they want, and so do the United States firms." Because of a short growing season, the large size of farms, and recurrent labour shortages, prairie farmers wanted equipment that offered speed. They also wanted accessible repairs and service which U.S. firms supplied through an extensive dealer network on the Prairies.[14]

Depression, War, and Postwar Changes

The Depression was devastating for the Canadian farm equipment industry. Sales dropped from about $41 million in 1929 to about $5 million in 1933, a greater decline than in any other industry in Canada. Imports dropped at an even greater rate. The Canadian government under Prime Minister R.B. Bennett, pursuing a protectionist policy in an effort to end the Depression, increased the tariff to 25 percent, although tractors valued at under $1,400 remained duty free. This did little to improve conditions for Canadian manufacturers, whose problems were compounded by an almost total collapse of export markets. With a change in government, the tariff was reduced in two stages to 7.5 percent by 1936. Poor agricultural prices, however, ensured that the sales of farm equipment remained low; in 1939, factory production had risen over the decade but still stood at only half of the 1929 level.[15]

World War II brought economic recovery to Canadian implement manufacturers. Although machinery was rationed in order to guide industrial production to meet war needs, farm equipment manufacturers were able to expand because they received government contracts for production of war-related goods. By the end of the war, the industry was strong, and with the end of rationing in 1945 and the removal of price controls in 1947, it was in a position to meet 15 years of pent-up demand for equipment. Rising farm prices further stimulated the industry, and in the years immediately following World War II, it was more profitable than most other industries. This was a revealing example of the nature of the farm equipment industry: in bad times it suffered more than most industries, and in good times it fared better. The postwar boom for farm equipment ended in 1953–54 when sales fell sharply, reflecting lower agricultural prices and the satisfaction of pent-up demand.

Despite high sales after World War II, in some respects the Canadian farm equipment industry was dealing with new conditions in the postwar years. In 1944, the import tariff on farm machinery was abolished, along with duties on materials and equipment for its manufacture. This created free trade with the United States, and an accompanying trend was that Canadian firms abandoned export markets other than the U.S. In part the consequence of currency controls applied by many countries in connection with postwar economic recovery policies, this trend also resulted from the creation of a single market in North America. While Canadian exports to the U.S. had been relatively low before 1939, by the mid 1950s they accounted for over three-quarters of Canadian farm equipment exports.[16] North America was now a single market for farm machinery. While full line firms had often integrated their Canadian and American operations before 1944, and indeed sometimes their world operations as well, total integration of North American operations became the norm under free trade.[17] By the mid 1950s, both Massey-Harris and Cockshutt, Canada's full line companies, had production facilities in the U.S., and the American giants, IHC and John Deere, had the same in Canada. Products being "sold throughout North America by the four companies are made in Canada or the United States, but not in both countries. The interchange of products is an important function in the Canadian-U.S. farm machinery trade."[18]

Clearly, the impact of free trade in 1944 was significant in the organization of the farm equipment industry. Imports also increased steadily and outpaced domestic production substantially; they rose from about 61 percent of total sales in 1945 to about 82 percent in 1954. Although tractors accounted for the greatest part of these imports, even when they are removed from the total values, a substantial increase in the import of other equipment is apparent. At the same time, exports from Canada increased from about 40 percent of production in 1939 to about 50 percent by the mid 1950s. While this was a substantial increase, it was still small in comparison to the increase in the value of imports. By 1954, almost all of Canada's exports went to the United States, and imports to Canada came almost exclusively from the U.S.[19]

Throughout these changes, the Canadian industry continued to be dominated by Cockshutt, IHC, and Massey-Harris (renamed Massey-Harris-Ferguson in 1953 to reflect its acquisition of Ferguson Tractor). These three firms accounted for about 90 percent of the Canadian industry's output by the mid 1950s. Massey had about 50 percent of this production, IHC about 25 percent, and Cockshutt about 15 percent. All of these firms continued to be located in Ontario, all within a 50-mile radius of Hamilton, the major steel centre in Canada.[20]

Consistent with its history of consolidation, the farm equipment industry continued to experience mergers and buy-outs. In 1961, Cockshutt was taken over by the White Group, which by 1965 had become a large conglomerate including Cockshutt, Oliver, and Minneapolis-Moline. By this point, the Canadian industry was dominated by IHC of Canada, Massey-Ferguson (by then probably the largest farm equipment manufacturer in the world), and John Deere.[21] Although mergers and consolidations continued into the 1980s, they no longer seemed to be sufficient to sustain the full line companies, and a number of long-established brands and companies disappeared. Massey-Ferguson was in difficulty from the late 1970s, and in 1986 it was restructured; in 1988, it went into receivership. White Motors Corporation of Canada was in receivership in 1980, and IHC's farm machinery division was bought by J.I. Case in 1984. The problems that the full line companies faced were variable and often unique to each company, but by the 1980s, most were in crisis. Many full line companies had not upgraded their plants to modern manufacturing systems quickly enough, and sales of large equipment like tractors and combines were declining as a result of poor agricultural markets and decreasing number of farms. In other cases, take-overs sapped company strength and capital, making them vulnerable in the notoriously cyclical farm equipment business.[22]

The increase in imports of farm equipment into Canada in the early 1950s continued into the 1970s. By 1986, Canada imported 93 percent of its farm equipment, and exported 79 percent of its production. The balance of payment deficit was about $1.3 billion.[23] The United States continued to supply most of these imports, and there were a number of reasons for the early and continued American domination of the North American market. For one, higher investment in specialized equipment in the 1950s gave American plants higher productivity per worker—even though their wage rates were higher than in Canada—and this created greater and more valuable output, especially of tractors. Of even greater importance was the efficient dealer network that U.S. firms had in place. By contrast, Canadian dealerships in the U.S. were poorly developed overall. And, since the American domestic market was much larger than Canada's, "it was natural for the large companies to emerge in the United States and to concentrate their plants in that country, where the greatest market existed." Free trade confirmed these tendencies.[24]

Competition and Sale of Farm Equipment

After 1915, there were a number of government inquiries and commissions that explored the pricing of farm equipment in Canada. Farmers argued that farm equipment cost too much, especially given the prices they received for their products. These inquiries generally condemned the pricing practices of the industry. For instance, while Massey-Harris cut prices by 10 percent in 1932, this was not a typical response by manufacturers to declining consumption during the Depression. As the federal inquiry in 1936–37 concluded, between 1891 and 1936 "retail prices of farm implements have been maintained at too high a level as shown by the financial returns of the companies engaged in the industry." Thus, further price increases were not justifiable.[25] Yet, while the index of machinery prices was relatively stable from 1920 to 1935, farm prices were extremely volatile, dropping in the early 1930s to less than a third of their level in the 1920s. Farmers argued that the stability of equipment prices compared with volatile commodity prices proved the existence of a conspiracy among the major companies to hold up prices. This was a conclusion the Saskatchewan attorney general endorsed in 1939 when he remarked that "however it was accomplished . . . there could be no denying that a combine existed to fix the price of farm machinery." Nonetheless, sufficient evidence could not be found for prosecution.[26]

These arguments about price rigidity were repeated after 1947 when wartime price controls were removed.[27] However, while a lack of competition on price characterized the industry before and immediately after World War II, the early 1950s saw a dramatic change in this practice on the Prairies. In 1953–54, the farm equipment market on the Prairies collapsed. Inventories were very high and losses on prairie operations were severe. But instead of waiting it out and holding prices up—the way it had been done in the past—competition on prices emerged in western Canada. In 1956, it was observed that there was little publicity about this price cutting and thus

> there is little general realization of the significance of what has happened. That is, the 50 or 60 year old agreement on prices, implicitly observed and rigidly enforced by all companies, went completely by the board and its place taken by a price competition of such pronounced character that no one, a few years ago, would have believed possible.[28]

Price cutting continued for the remainder of the 1950s on the Prairies, although it was apparent by 1959 that it was drawing to a close. By 1970, the old pattern had fundamentally been restored.[29]

Although high prices were most often blamed on a lack of price competition due to collusion, competition in other areas was said to contribute unnecessarily to high costs. Competition in the development of new technology and equipment was often fierce among the industry giants. Farmers were frequently critical of this practice and claimed that "luxuries" and "frills" designed merely to attract purchasers forced up prices. These claims were often overstated, however, and defining a frill was sometimes difficult—for example, were soft seats on a tractor a frill or a necessity? [30] Nevertheless, technological change was a key to increased sales and was used to attract buyers. It did, however, have its drawbacks and forced manufacturers to plan carefully. It remained as true in the twentieth century as it was in 1893 when *The Nor'-West Farmer* commented that "the life of the manufacturer of farm machinery is not a bed of roses. Improvements follow each other with such wonderful rapidity that old implements soon become useless and the manufacturer who is not always on the alert finds himself loaded up with a stock that is out of date and unsalable."[31] Additional problems with inventory were inevitable in a cyclical industry. Production had to be scheduled well in advance of sales, yet the whole market could collapse because of a sudden downturn in agricultural prices. It was not unusual for manufacturers to be forced to carry a large inventory until the market improved.[32]

Mass production could minimize the costs of continuously changing technology and style, but these cost savings (as enjoyed by the auto industry, for example) could rarely be fully realized, even by full line companies with large markets. Production runs of farm equipment were too small and the number of models and variety of equipment were too great. Additionally, the design of farm equipment was far more complex than, for instance, the automobile. While still complicated, automobile engineering required only that it function on surfaces built especially for it. In contrast, a tractor, for example, had to be designed to tow or push implements as well as work with highly variable conditions and surfaces.[33]

This constantly changing technology grew from new needs in agriculture, the desire by manufacturers to present new equipment to earn sales, and farmers' demand for new equipment for functional and status reasons. While this continual change increased prices, an additional factor was distribution, another area of fierce competition among the industry giants. Although the 1936–37 House of Commons inquiry noted that there was no evidence of factory inefficiency, it argued that the costs of distribution were too high. The huge dealer network maintained throughout the Prairies increased manufacturers' overhead, and the practices encouraged by this system were costly. It reported that "there is competition in the matter of securing sales but little or no effective competition in the matter of prices as between companies." Retail prices were set by the manufacturer, not the dealer, and price increases occurred simultaneously among all companies, an event the Committee found a "remarkable coincidence."[34] Competition was thus on nonprice elements, or as one commentator phrased it in 1944, "keen competition . . . forces uneconomical practices in servicing, in trade-in policy and in extension of credit—all of which, in the final analysis, the farmer must pay for, in part at least, in the price of the product."[35]

Competition Through Distribution

While their dealer networks gave American full line firms a distribution advantage in Canada, dealerships were central in the operations of all full line companies. In the late nineteenth century, many of the large companies realized that their dealer networks were essential to success and began to take greater care in their selection and strategic placement.[36] Similarly, they quickly established wholesale operations in emerging markets like the Canadian west. For example, John Deere opened a wholesale house in Winnipeg in 1878, followed by A. Harris and Sons in 1879 and Cockshutt in 1892. Now poised to take advantage of the new market, they were also

Because of their size and organization, full line companies made an early appearance on the Prairies. Although Minitonas, Manitoba, was only a tent town in 1899, it already had a Massey-Harris dealer. (Minitonas 1, PAM)

sufficiently well capitalized to survive periods of slow growth and wait for boom times.[37]

Throughout their history, full line companies integrated manufacturing, distribution, and sales. Each company employed a "blockman," a sales representative responsible for a limited number of dealers. Blockmen were important in co-ordinating the whole structure of integrated manufacturing and sales, which was based on the local dealer. In most towns, each firm had a dealer who sold only their machines and repair parts, provided servicing, and arranged financing for purchasers. He exclusively represented the company in the community and was the vital link between farmer and manufacturer. This system was crucial because of the nonprice competition that characterized the industry and because it tied directly to farmers' motives for purchasing particular brands of equipment. While earlier evidence is scattered, a 1970 survey of farmers seems to reflect general attitudes to the purchase of farm equipment, attitudes that had persisted from an early date. The survey showed that a majority of farmers bought particular brands of equipment because they were familiar with the brand's performance and because of the dealer's reputation for providing service and stocking repair parts. Equally important were informal conversations with neighbours and the observation of equipment on nearby farms. Trade-in allowances for old equipment were also a significant part of attracting business.[38]

Given this framework, a network of stable dealers throughout the country was essential. This also meant that advertising was of secondary value in direct sales, although it was significant in keeping a brand name in the public eye. While all full line companies placed advertising in newspapers and magazines, other promotional

Displays of farm equipment always attracted the public and created opportunities for manufacturers to mount community promotions of equipment. This one was held in Stettler, Alberta, in 1910. (Archives Collection, A6098, PAA)

efforts were also used. As in the United States, late nineteenth century prairie dealers mounted special events called "dealer days" to display and demonstrate equipment. For a time, some companies also used special trains to carry displays of equipment around the country, but more often they and their dealers mounted displays and demonstrations at local fairs and exhibitions.[39]

The larger summer fairs always featured farm machinery, as in Winnipeg in 1911. (Foote Collection, 1484/N2420 PAM)

The relationship between dealer and farmer was not always an amicable one. While farmers usually welcomed new machinery, they were often critical of dealers. Some farmers claimed that dealers persuaded them to purchase machinery that they did not need, that was not mechanically sound, or that did not deliver the results promised. In part, this attitude may have been reflected in the periodic assertion by farmers that all retailers charged excessive prices and were merely "middlemen" who skimmed profits off sales. Yet, commissions charged by farm implement dealers seem not to have been excessive before World War II, although their business investment was relatively low since machinery was carried on consignment.[40] But the dealer was also wedged in by an often unsatisfactory relationship with the manufacturer. As one dealer from Unity, Saskatchewan, remarked in 1939, he was "the buffer between the farmer and the company, a man who does business without any protection."[41]

In an attempt to strengthen their hand, some dealers moved to form associations to represent their interests. Wholesalers had successfully done so by the late 1880s in Manitoba, but this rested on the dominance of the wholesale trade by full line companies, which created a tight and coherent group.[42] In contrast, retailers initially were less successful. One such attempt was the formation of the Retail Implement Dealers Association in 1908 in Manitoba which aimed to create "a fraternal feeling among the implement dealers." The association tried to regulate competition, come to "an amicable understanding as to terms and prices," and regulate the extension of credit. In this format, the organization found little success. In the late 1920s, resentment of the pressure placed upon them by the full line companies led prairie dealers to begin once again to organize. The most successful of these organizations were at a district level, and by the late 1930s, they were beginning to coalesce into provincial organizations. The first successful regional organization of prairie dealers came in 1945. These organizations generally focused on lobbying for their sector of the retail trade rather than on controlling conditions of business. The prospects for effective organization of dealers on such terms was limited in any event: they were tightly controlled by the full line companies, and there were too many dealerships in most towns. As well, by the late 1930s, they were suffering from mail order catalogue competition on parts, their most lucrative merchandise.[43]

After World War II, the dealer system changed. Full line dealers no longer took a commission on consigned equipment. Except for those representing John Deere, dealers had to purchase equipment on short-term credit arrangements from the manufacturer. For the dealer, this considerably increased both risk and potential for greater profit. This change played a part in destabilizing the price maintenance that had characterized the industry for so long. With rapidly falling sales in 1953–54, dealers independently reduced their profit margins to garner sales, further depressing prices.[44] After this market collapse, manufacturers were forced to give easier credit to dealers as an incentive to maintain sufficient inventory and to keep equipment on display. This was the beginning of what was called "floor planning." Manufacturers provided their dealers with an interest-free loan for up to 23 months for the purchase of equipment, except for tractors which were interest free for 12 months. It was an expensive method of merchandising, adding perhaps 5 to 7 percent to the cost of equipment.[45]

Dealers played an important role in equipment warranties and trade-ins. The manufacturer paid only for factory defects, while all other adjustments were made by the dealer. Initially, dealers alone took responsibility for trade-ins, although by the 1950s, the floor planning concept was used for trade-ins as well as new equipment. Technological change was the basic incentive for trade-ins. When equipment was purchased because of its new features or larger size, the farmer wanted to trade in the equipment he already had. While this was significant in the interwar years with tractors and combines, from 1946 until 1951 trade-ins were unusual. Demand was so high that farmers themselves could easily sell their old equipment and dealers could simply refuse to take it without damaging sales. However, trade-ins soon increased as agriculture further adjusted to mechanization and the pent-up demand for machinery was met. Farmers were reluctant to deal if they could not trade in old equipment. Yet, overall, there was little profit for dealers on the resale of trade-ins, although they were often essential in doing business. The resale of trade-ins was complicated by the fact that there were no standard values for such equipment, and there were a number of attempts to address this problem. For example, in the late 1930s, the Manitoba Retail Implement Dealers' Association drew up a list of average prices paid for tractors. While it was anticipated that the list would be a "valuable asset" for dealers, it was probably of limited benefit given the level of competition among companies.[46]

The dealer was also the link between the manufacturer and the farmer for credit. Even in good times, few farmers could pay cash and most needed financing for at least half of the purchase price. Since finance companies and banks did not lend money for the purchase of farm machinery before World War II, financing of equipment purchases was provided almost entirely by manufacturers. This not only increased their capital needs considerably but also meant a high level of assets in the form of accounts receivable. Nonetheless, the companies gave only a small discount for cash, increasing the attraction of credit buying. Interest rates charged by the companies were high (9 percent in the early 1920s), and they were obviously set to allow considerable leeway for losses. There was considerable buying of farm machinery in good times and when credit was easy, and losses on loans were high. These losses were ultimately passed on to consumers, and it was claimed in 1937 that one manufacturer alone had losses in Western Canada that amounted, over the previous 15 years, to 4 percent of total sales, or $2.75 million.[47]

Farm-equipment financing changed after World War II with the passage of the federal *Farm Improvement Loans Act* in 1944. Under this legislation, farmers could obtain low interest loans for, among other things, farm machinery. In 1952, almost $91 million in loans were made, three-quarters of which was extended to prairie farmers. It was a highly significant initiative and helped both to sustain the postwar expansion of farm mechanization and to meet the pent-up demand for equipment. As well, by the 1960s, many finance companies had begun to lend funds for machinery purchases.[48] These developments gave purchasers some flexibility since financing no longer depended solely on the manufacturer. While they continued to be important sources of credit, they were not the only one, and this increased availability of credit was important in increasing sales. As one dealer claimed in 1952, the *Farm Improvement Loans Act* had "saved 75 percent of us [dealers] from bankruptcy."[49] This was especially significant to dealers because manufacturers were reducing their dealer networks at this time. Massey-Harris, for example, had 1,957 dealers in Canada in 1944 but only 754 by 1966. The company also closed some of its distributing centres. While this was an ongoing phenomenon after World War II, the succession of company consolidations in the 1970s and 1980s significantly reduced the number of dealers and increased the territory of those remaining.[50]

The full line companies typically displayed the size of their factories in their promotional literature in an effort to stress their success and permanence. This illustration appeared in the Massey-Harris catalogue, 1898.

The farm equipment industry was not unique in facing a highly cyclical and often unpredictable market. But the complexity and cost of its products created a unique set of problems in scheduling production and inventory control. These problems were compounded by the importance of technological change and by distribution as the basis of competition between manufacturers. The structure built by the full line manufacturers, which featured an integrated hierarchy of manufacturing, distribution, sales, and credit, created a rigid and often bureaucratic system. Yet it was able to accommodate the nature of the North American farm implement market, simultaneously establishing the standards of the industry and shaping the farmers' views of what was desirable in machinery retailing.

Farmers and the Full Line Industry

The ability of the full line companies to set both the standards of the industry and the framework within which all manufacturers of farm implements operated was a clear indication of their dominant role in the farm implement market. Prairie farmers argued that the marketing strategies used by the full line companies increased prices and ignored the best interests of consumers. Such claims had several effects on the operation of the industry on the Prairies. Provincial governments tended to accept the validity of many of these charges and passed legislation regulating the industry. For the same reasons, they also became involved in the testing of farm implements. Some farmers also took economic action by establishing co-operative distribution and manufacturing of equipment, especially through Canadian Co-operative Implements Ltd. (CCIL) of Winnipeg.

Regulating Farm Machinery Sales

The earliest issue tackled by government concerned warranties and sales ethics. Farmers contended that dealers made promises about performance that were untrue, that manufacturers refused responsibility for the representations of their dealers, that purchasers were forced to sign contracts that were unfair, and that repair parts were not readily available or were no longer produced. They also charged that credit terms were unfair. While such complaints were rarely made in respect to small implements like plows, binders, and harrows, they were commonly made about power equipment like tractors and steam engines.[1] It would seem that many of these early complaints were justified. During the settlement years, there were a number of "fly-by-night" dealers who sold useless or unnecessary equipment to unsuspecting and uninformed settlers.[2] This reputation stayed with the industry, although by the interwar years, it was increasingly undeserved.

In response, all Prairie Provinces enacted legislation to regulate the farm equipment trade. Demand for this legislation was loudest from farmers, but it was not exclusively their concern. Alberta's 1913 legislation, for example, was vigorously supported by the Associated Boards of Trade of southern Alberta, and parts of their draft proposal were reflected word for word in the legislation.[3] The Alberta legislation governed all farm implements and machines and overrode any unreasonable provisions of contracts between vendor and purchaser. It forced the manufacturer to take responsibility for claims made by its agents or dealers and stipulated that warranties had to be upheld "notwithstanding anything in any agreement contained to the contrary." [4] Similar provisions appeared in legislation enacted in Saskatchewan in 1915 and Manitoba in 1919, although Manitoba's legislation was more extensive. It regulated prices charged for repairs and required manufacturers to file with the province an annual list of prices and changes in the company's credit policy.[5]

Except for the addition of provisions requiring that repair parts be available for a set period after the date of sale, this legislation basically governed the sale of farm equipment on the Prairies until after World War II. By the late 1940s, however, major revisions began, although this seems to have been more a reflection of farmer frustration with rising machinery prices rather than a decline in the quality of machines. In 1949, the Saskatchewan legislation was amended to provide, among other things, that all implement dealers be licensed. The most extensive reworking of its legislation, however, came in 1958 with the creation of an integrated administrative structure called the Agricultural Machinery Administration. It was governed by a special board and had the power to administer the *Agricultural Machinery Act*, test equipment, and carry out developmental work on farm equipment. Earlier provisions requiring stocking of repair parts and licensing of dealers were retained, but a new provision required that manufacturers selling farm equipment in Saskatchewan had to establish a distributorship in the province to carry spare parts. Provincial manufacturers could act as their own distributors.[6] While this provision was unique among prairie implement legislation, it shared with the other provinces' legislation the characteristic of putting responsibility for repair parts on the manufacturer, not the dealer. It was primarily designed to quiet widespread complaints about the availability of parts, but it had a seemingly unexpected impact on the economy. Most distributors were located in Winnipeg, and many closed these operations and moved to Regina rather than maintain two prairie distribution points. In conjunction with Regina's locational advantages and the size of the Saskatchewan farm equipment market, this made the city a primary parts distribution point in the region.[7]

A decade later, Saskatchewan again amended its legislation. Most existing provisions were retained, but a new and highly controversial one was introduced. A fund that was to be financed through a levy on distributors was set up for compensation of farmers suffering loss from delays due to unavailability of repair parts. The government defended this action by arguing that complaints by farmers about warranties and supply of repair parts had increased.[8] The provision came into effect in Saskatchewan in 1973 with the establishment of the Agricultural Implements Board, the administrative agency that adjudicated applications for compensation. By this time, all three provinces required that repair parts be available for 10 years after sale of equipment, as well as on an emergency basis. Manitoba further required that all new combines and tractors sold in the province carry a three-year warranty, including all component parts, regardless of manufacturer. Other equipment had to be warranted for one year.

The farm equipment industry generally opposed this legislation. While it had always objected to regulation that singled out farm equipment, some believed that particular provisions had gone too far. Joe MacDonald of MacDon Industries, a Winnipeg short line manufacturer, accurately pointed out that no other industry was required to operate under such conditions. However, some of these requirements were not unique to Canada. In 1965, the Farm Equipment Manufacturers Association, a United States industry association, established a policy that its members, among other things, maintain stock or production of repair parts for at least seven years after sale of equipment. Another industry complaint focused on its need to meet separate regulations in the three provinces even though they essentially formed a single farm equipment market. By the 1970s, the three provinces were addressing this concern through greater informal co-operation on uniformity in their regulations.[9]

Testing of Farm Equipment

A related issue concerned testing the performance and durability of equipment. All full line companies extensively tested their equipment. Since they were marketing equipment for the whole of North America, equipment design aimed to meet the broadest possible range of conditions. For smaller manufacturers, elaborate testing was impractical, but here too equipment was usually tested as extensively as possible before marketing. Normally, most equipment was tested for two to four seasons, and redesign was undertaken as weakness was revealed.[10] Regardless of the level of testing, most equipment was not perfect for all conditions when first introduced. Many farmers resented this, claiming that they were being used as "guinea pigs" by the implement companies. This was no doubt true in some cases, but not overall. As John Brown of CCIL observed, poor design cost a company money and damaged the reputation of its engineers, management, and products.[11]

While testing was never mandatory under any of the three provinces' farm implement legislation, evaluation of the suitability of equipment for prairie conditions had been carried out from the early 1920s by the federal government experimental farm at Swift Current. The objective was "to thoroughly try out new machines" to help devise economical use of equipment and adapt it for dry land conditions. The findings were published in the station's annual reports or in special publications. While it was an aspect of the station's educational work, it also assisted manufacturers in developing equipment. For example, the first combines on the Prairies were tested at Swift Current in the 1920s, and C.S. Noble, who invented special tillage equipment for dry land farming in southern Alberta, had his machines tested at various experimental stations. Similar tests were carried out on other equipment of local design. The same type of work was done by most prairie universities.[12]

Testing before 1945 was also promoted by the Agricultural Engineering Committee, organized under the *Prairie Farm Rehabilitation Act* in 1935. The committee acted as an advisory body to government on agricultural implements and their relationship to soil conservation. In later years, this focus broadened to include a range of extension and educational issues concerning farm equipment. The committee was dominated by agricultural engineers from government and universities, and although it did not directly carry out testing, it was an important force in standardizing and co-ordinating research and policy on farm equipment. As the committee noted in 1936, while universities and experimental stations did significant work in equipment testing, these efforts were unco-ordinated and secondary to other research activities. By urging that equipment testing be given a rightful place in agricultural research and publishing, the committee had played a significant role in legitimizing practical equipment testing on the Prairies by the 1940s.[13]

Public testing had always been a concern, but after World War II, those advocating the testing of new equipment gained ground. This was not the result of a sudden decline in the quality of machinery; indeed, most observers—even the most critical—admitted that design and fabrication of farm machinery was better than it had ever been.[14] Instead, as in the case of regulation of equipment sales, the steadily increasing cost of equipment made farmers ever more conscious of getting value for money. By 1962, *Canadian Farm Implements,* a trade publication, noted that while manufacturers were competing more strenuously through model changes and redesign of equipment, farmers were under "intense pressures of rising costs" and were "thus sensitive and increasingly militant in the market."[15]

In this context, revisions to Saskatchewan's legislation in 1949 included provisions for the appointment of an Implement Engineering Committee to test the suitability for provincial conditions of equipment and repair parts sold in Saskatchewan. These provisions, however, were not acted upon until 1958 when the province established the Agricultural Machinery Administration (AMA), a body set up to administer the legislation and to test equipment. It was the first government testing agency in Canada. Laboratory and field tests were used to simulate farm conditions, and weaknesses exposed during this process were reported to manufacturers. The AMA issued pamphlets reporting the test results, recommendations for improvement, and the reply of the manufacturer to the comments. By the early 1960s, about 15,000 copies of each pamphlet were distributed in Saskatchewan alone, with many more sent to Alberta and Manitoba. Equipment was selected for testing if it was used extensively in the province, or if it was relatively new or had a history of structural and functional defects. The AMA's small staff limited its programs, although in 1960–61, it conducted tests on 20 machines. It also tested, on a fee-for-service basis, prototypes for manufacturers. These tests provided the manufacturer with information about the implement's performance under Saskatchewan conditions. In all cases, tests were comparative—each machine was compared with a "standard" piece of equipment selected on the basis of its popularity in Saskatchewan.[16] The greatest benefit of this testing accrued to consumers, but it also benefited manufacturers. Impartial and public test reports, if favourable, had substantial commercial value in marketing, even though they could not be used in advertisements. This was also valuable for export markets where objective test results assisted in gaining acceptance for largely unknown manufacturers. In addition, all tests, because they were controlled, provided objective information for design engineers. This was potentially useful to small firms with limited engineering budgets.[17]

The AMA received considerable support from farm organizations, but less from the farm equipment industry. While the tests were in effect a contribution to a company's research and development, many companies were critical of the AMA, perhaps fearing that a negative test report would hamper sales.[18] But it was also true that testing criteria on what constituted "reasonable use" were imprecise, which could lead to an unfavourable, and not wholly justified, report.[19] In 1964, the newly elected Liberal government in Saskatchewan, acting on its long-standing opposition to government testing of farm equipment, abolished the AMA's testing activities on the grounds that testing was not a job for government and that it duplicated tests done by manufacturers. The new government was also concerned that although the other Prairie Provinces contributed little or nothing to the program, they benefited from its work.[20] The testing equipment and some staff were transferred to the University of Saskatchewan; this represented a shift in emphasis in testing procedures because the university concentrated on studying the function and impact of equipment, not the testing of specific brands. It did, however, carry out specific tests relating to design development on a fee-for-service basis for manufacturers.[21]

There was considerable criticism of the government for abolishing the AMA's testing, and public pressure for its re-establishment endured. This was bolstered in 1967 by the Barber Royal Commission pronouncement that the AMA had been valuable and should be reinstated.[22] This debate indicated that testing had become a political issue in Saskatchewan, and it was therefore not surprising when the New Democratic Party, after its election in 1971, announced plans to re-establish a farm equipment testing facility, ideally on a joint basis among the three Prairie Provinces.[23] As the Barber commission had observed, a proper testing facility needed a larger budget than the AMA had enjoyed, and it was unrealistic to have a single province bear the full costs of what essentially was of regional benefit. The Saskatchewan government accepted this view and invited the governments of Alberta and Manitoba to join in establishing a joint testing facility.[24]

The Prairie Agricultural Machinery Institute (PAMI), the new testing facility that grew out of these discussions, was set up in 1974. It was based in Humboldt, with a research branch at Lethbridge, and the three Prairie Provinces shared the costs. In 1977, another branch was established in Portage la Prairie. PAMI used approximately the same testing approach as had the AMA. In 1988, Alberta withdrew from PAMI to operate the Lethbridge facility as The Alberta Farm Machinery Research Centre, leaving Manitoba and Saskatchewan to fund PAMI.

Six hundred pieces of equipment had been evaluated by PAMI by 1989. Like the University of Saskatchewan and the earlier AMA, PAMI also tested prototypes of equipment and worked with manufacturers to modify designs. It also helped inventors by testing their ideas and referred promising ones to appropriate government departments for assistance with securing patents. Further, it carried out applied research on mechanization and cultural practices. In this way, it directly contributed to prairie agricultural engineering.[25] As with previous testing, there was some criticism of PAMI within the industry, but most viewed it as a positive development. The Prairie Implement Manufacturers Association, the prairie industry organization, also supported PAMI as a valuable vehicle for research, development, and testing of equipment, which, it reasoned, would strengthen the industry on the Prairies.[26]

Challenging the Structures of Full Line Equipment Manufacturers

Testing of equipment and regulation of sales practices were only two aspects of the public debate about farm equipment. How to bring the costs of equipment into line with farmers' incomes was also of great concern. Throughout, various solutions for cost reduction were proposed. The most obvious was for farmers to take better care of equipment and extend its useful life. Some observers contended that prairie farmers were especially hard on equipment; in 1902, for example, the Prairies were called "the greatest machine butchery in the world." The need to educate farmers to care for their equipment long remained a refrain of governments, educational agencies, and the farm press, all of which offered a steady stream of advice to farmers on machinery care.[27] This self-help approach also held that farmers should share equipment to reduce costs. Widely practised on an informal basis for some types of equipment, the enduring popularity of this approach was still evident during the 1960s in federal government loans schemes for farmers who formed "syndicates" to purchase and pool use of equipment. A number of farmers took advantage of these programs, although their overall popularity was low.[28]

Proposals for mass buying, in which a government agency purchased and then resold equipment, were also popular. The Social Credit government of Alberta implemented this approach in 1939, selling farm machinery repair parts through its provincial marketing agency. There was, however, little that provincial governments could do about pricing since any practical solutions lay outside their jurisdiction.[29] Others recommended that dealers standardize trade-in values and policies or that trade-ins be handled by dealers selling only second-hand equipment. Still others argued that floor planning should be abolished to reduce distribution costs and that dealer dependency on a single manufacturer be eliminated to allow more competition.[30]

Before being removed in 1944, tariff reduction as a means to obtain lower prices was proposed more often than any other solution. Farmers habitually blamed the tariff for increasing machinery costs and restricting prairie agricultural development.[31] By 1914, it had already become a cliché in the region to describe the tariff as "a monster that increases the price of farm machinery." [32] Commonly, it was argued that Canadian farmers paid more for equipment than did Americans because of the import tariff. But when the tariff was reduced by almost 20 percent in 1935 and 1936, Canadian prices declined only slightly, and in some cases actually increased. Given the nonprice competition that characterized the industry at the time, the full line companies found no need to reduce prices to maintain market share. Although demands for tariff reduction continued until 1944 when the tariff was removed, they were largely pro forma. Tariff reduction was obviously not the solution, giving credence to earlier ideas that farmer-controlled co-ops that distributed and manufactured farm equipment were the answer.

Co-operative Manufacturing and Distribution

The idea of co-operative distribution of farm machinery was not unique to the 1940s or to Canada. It had been used in the United States—among the earliest being in Illinois in 1905—and had inspired calls for a similar arrangement in the Canadian west.[33] The first formal recommendation of this approach was made in the 1915 report of the Saskatchewan Royal Commission on Farm Machinery.[34] By this point, the United Grain Growers (UGG), a co-operative elevator company, was already buying farm equipment in bulk and reselling it in an effort to lower machinery costs. Beginning in 1914, the UGG sold equipment for cash by mail order, but it found this unsatisfactory and went to an agency approach. This was little better: supply of repair parts was poor, it did not accommodate the cyclical nature of the market, and customers' needs (or demands) for credit could not be satisfactorily met. Despite these problems, the UGG experienced "a good deal of agitation" to purchase a machinery manufacturing plant in the U.S. or Canada. In light of its difficulties in selling farm equipment, the company refused to take up this challenge, and in 1922, it abandoned the farm equipment market entirely.[35]

The idea of distributing farm equipment co-operatively, however, was not dead. In 1939, the Saskatchewan Royal Commission on farm equipment recommended the formation of a distributing co-operative for farm equipment. In addition, the commissioners proposed that a campaign be undertaken to educate farmers about machinery prices. This campaign would operate in conjunction with the organization of local co-operative associations to distribute implements supplied by a central co-operative organization. This was a radical departure. While co-operative systems were not unusual on the Prairies, the analysis of what stimulated high farm equipment costs reflected in this particular proposal was a significant change.

In an effort to force them (or government) to reduce prices, farmers had traditionally attempted to show that the machine companies were "unconscionable robbers." However, as John Brown, the president of the resulting co-operative observed with a blunt honesty, "few indeed ever appeared to wonder if the farmer did not share some of the responsibility for high prices." Farmers demanded a service, distribution, and marketing system that was expensive. As Brown noted, no proof had been found that farm equipment profits were higher than those of other industries. Yet, while competition in manufacturing created efficiency, "competition in the distribution of most goods, while forcing efficiency in the debasing techniques of salesmanship, results in wastes, in senseless expenses which go a long way in discounting the economies of competitive production." Evidently, an educational campaign was required to persuade farmers to help themselves. It was clear that "the machinery buying habits of farmers were deeply ingrained and that changing those habits would take a good deal of time." This was, however, "the all-important problem" that the co-operative had to solve.[36]

Along with changing farmers' attitudes, the co-operative planned to obtain equipment on a cost-plus basis from a Canadian full line manufacturer and resell it through co-operative outlets. It was further hoped that the co-operative might handle the complete distribution of that company's products through its co-operating outlets. Saskatchewan immediately enlisted the support of the other two Prairie Provinces and prairie co-operative organizations like the wheat pools to incorporate a company for these purposes. A campaign to enlist farmer support also began, and farmers were asked to buy a minimum of $10 and a maximum of $40 in shares in district implement associations. By 1940, sufficient subscriptions had been signed up, and a company called Canadian Co-operative Implements Limited (CCIL) was formed.[37]

CCIL was not immediately active because wartime controls made an agreement with a Canadian manufacturer impossible. Nevertheless, subscriptions continued to be sold and policy developed. In 1942, a major shift in focus took place when it was decided that CCIL would not only distribute but also manufacture equipment. In 1944, the three Prairie Provinces, the western wheat marketing co-operatives (such as the wheat pools), and consumer co-ops in the three provinces agreed to push ahead by recruiting additional members, seeking an agreement with a Canadian manufacturer to supply equipment, and borrowing funds to purchase a factory. The three provinces lent a total of about $250,000 to CCIL, which purchased the Winnipeg plant of Gregg Manufacturing, a short line manufacturer. Additional capital had been subscribed from farmers, bringing the total to almost $1 million. Over half the subscribed funds came from Saskatchewan.

Like most consumer co-ops, CCIL did not sell at reduced prices. It paid a yearly cash dividend based on purchases, retaining some dividends for the business's capital. It sold only for cash. When the company was formed, it was anticipated that the district associations would become "the nucleus of a district machinery and supply depot." This would overcome one of the reasons for increased farm equipment costs, namely, the expense of "a multitude of small dealers carrying limited stocks of implements and repairs but without adequate service facilities."[38] The company was in reality an amalgam of district associations. In 1948, this organizational structure was revised: the district associations were dissolved and a new entity (still called CCIL) was chartered. Shares held in the district associations were exchanged for shares in CCIL. This created a more unified and focused company, but its democratic nature was preserved through the use of elected delegates to govern the company.[39]

The immediate and crucial task was to obtain an agreement with a manufacturer to supply complex equipment that CCIL could not manufacture. After lengthy negotiations, CCIL entered into a contract with Cockshutt in 1945. Cockshutt was to supply equipment to CCIL, as well as to the National Farm Machinery Co-operative Inc., an American farm equipment co-op. CCIL had joined National in 1942 in an earlier, but unsuccessful, effort to obtain tractors. Cockshutt planned to manufacture tractors, but it needed sales in the United States to make this production feasible. An agreement with National was a means of penetrating the U.S. market without having to bear the cost of establishing an American dealer network. The route to this market lay through CCIL: Cockshutt agreed to supply CCIL, but only if CCIL persuaded National to take Cockshutt equipment. CCIL's relationship with Cockshutt was thus secondary to Cockshutt's ambitions in the U.S. But the essential relationship between CCIL and Cockshutt had been established, although as John Brown recalled, "during all the association with Cockshutt, we occupied, at all times, the position of a petitioner. There was no possibility of getting such supplies elsewhere or of manufacturing them ourselves."[40]

CCIL's vulnerability was further demonstrated in 1949 when its agreement with Cockshutt was renegotiated. CCIL's cash dividends to members were now limited to 5 percent on Cockshutt lines, and retained dividends could not be used to purchase equipment from Cockshutt. In order to protect its existing dealer network, Cockshutt also restricted the locations at which CCIL could establish depots. Clearly, the relationship between CCIL and Cockshutt, although "reportedly good," was inherently unstable, even though by this time CCIL was buying tractors, combines, mowers, and swathers from Cockshutt.[41]

CCIL initially investigated numerous types of equipment for manufacture. Its aim was to develop new ideas rather than copy equipment already on the market.[42] It began to manufacture tillage equipment and attempted to create a market niche and farmer confidence in the company by providing equipment that would contribute to soil conservation and suit the new tillage practices coming into vogue. In this, CCIL was a strong innovator, bringing out new equipment that made significant technological contributions to prairie agriculture. The most important of these early innovations was the "Disker" in 1946–47. The company also manufactured other smaller equipment. Given its relative success, there was some discussion in the early 1950s that CCIL should expand the lines it manufactured. CCIL resisted these proposals, arguing that it did not welcome the risk associated with engineering more complicated equipment. It also noted that Cockshutt's equipment was satisfactory and that it did not want to assume the debt such expansion required. Tellingly, it pointed out that "manufacturing profits over the long run were not excessive," particularly in light of "the risks involved." [43]

At the close of the 1940s, CCIL was still attempting to lower the costs of farm equipment through controlling the economics of distribution. It continued to reject the traditions of the industry and avoided sales pressure on the grounds that the farmer should "buy machinery rather than wait to be sold."[44] The company used minimal advertising, downplayed features like showrooms, and kept its sales staff on salary or on low commissions. In an effort to respond to farmers' general complaints about testing and performance of farm equipment, questionnaires asking for suggestions for improvement were sent to purchasers of equipment each year. CCIL also continued to sell for cash and used patronage dividends to pass along savings to consumers. Similarly, it did not take trade-ins but sold used equipment on a consignment basis, in this way rejecting the use of the trade-in as a bargaining point to lower the purchase price of new equipment. This was consistent with the traditional argument of farmers that the trade-in system was in reality only a means of buying the farmer with "his own money."[45] In all, CCIL's approach was rather austere, although it was in keeping with what many prairie farmers had for half a century said that they wanted.

But CCIL soon had to confront the problem that farmer rhetoric about equipment manufacturing obscured important complexities, both about the industry and about the reality of farmers' behaviour in buying equipment. Because of cost, CCIL had not organized district depots as originally planned. Instead, established retail co-ops and wheat pool and UGG agents throughout the Prairies were used for local distribution. Servicing was handled through head office in Winnipeg or through branches in Saskatoon, Regina, Edmonton, and Calgary. The company also provided a mobile repair service. Panel trucks stocked with repair parts and fitted with equipment (such as welders) were used to service equipment. This overall approach was, however, quickly found to be unsatisfactory. Service was too slow and retail co-ops had neither the expertise nor the personnel for sales. CCIL was therefore forced to use depots to meet farmers' needs and wants. The first one was established in the Peace River country of Alberta in 1951 and others followed shortly in the rest of the Prairies. The depots were run by salaried staff. While CCIL was restricted in its choice of location by its contract with Cockshutt, it had opened 25 depots by 1952.[46] This contributed to a doubling of distribution costs between 1946–47 and 1951. Although the company resisted these changes and expenditures, as John Brown noted, it made them "out of the bitter necessity of having to exert a great deal of effort to persuade farmers to support the organization that had been set up to reduce prices of farm implements."[47]

CCIL was, in effect, shifting its direction to accommodate farmers' actual purchasing practices and their need for service. Obviously, hopes of leading farmers into new ways of purchasing equipment had not been realized. Old habits died hard, and the company found it difficult to overcome "the inertia of farmers with respect to brands of machines purchased and also as to buying habits." Yet CCIL recognized that if it attempted "to conform completely to traditional buying and selling habits then no economies could be realized."[48] Nevertheless, by the early 1950s, CCIL was finding it necessary to adopt many of the practices of its full line competitors. For example, in 1952, CCIL depots began accepting trade-ins, a move that added considerably to the company's administration and carrying costs.[49]

Many of the basic issues faced by CCIL in its first decade and a half continued to characterize the company's operations.[50] Initially, CCIL was relatively successful, with sales increasing from about $1.5 million in 1946 to about $5 million in 1951. Consistently, about one-half of these sales were in Saskatchewan, but they represented only about 5 percent of total prairie equipment sales.[51] This market share did not make CCIL a serious force in the farm equipment industry, nor did it enable CCIL to build sufficient cash reserves for hard times. Because of this, the collapse of the farm equipment market and the emergence of price cutting in 1953–54 hit it especially hard. And, the acceptance of trade-ins had created the added burden of carrying an inventory of used equipment.

CCIL was never able to persuade farmers that a co-operative approach could be in their interest. While most farmers showed no hostility to the principle behind CCIL, they displayed little enthusiasm for the company and were unwilling to adjust to its different approaches. CCIL's "dealer-customer" relations were inadequate, and it lacked "smart and efficient" depots. Farmers persisted in "shopping around" for equipment and using trade-ins as a bargaining ploy. They also showed a decided reluctance to abandon brand loyalty or to assess critically the role that equipment played as a status symbol in farm life.[52] Yet despite CCIL's overall minimal impact, throughout the 1950s and early 1960s it remained a machinery company like no other. Its management often held a pure profit motive in contempt and argued that "our concern was altogether different. It was to provide good machines that would help the farmer to farm more efficiently and also get his machines at lower cost." [53]

The 1960s were a decade of change for CCIL. The company was able to get back on its feet following the downturn of the mid 1950s, and by 1966, sales had reached $20 million. Yet this represented only 5.7 percent of total prairie sales of farm equipment, and the company seemed unable to improve on this minority share of prairie sales. Part of the problem was caused by unstable equipment supply. While some manufacturers refused to supply CCIL with equipment (arguing that CCIL damaged the dealer system and therefore threatened the industry), the first serious blow came in 1961 when White took over Cockshutt and cancelled its supply agreement with CCIL. This forced CCIL to find new suppliers of heavy equipment. It turned to Europe and imported Deutz and later Volvo tractors, and it also imported Kleis combines from Belgium. However, when New Holland bought Kleis, it refused to supply repair parts to CCIL, arguing that it planned to manufacture combines in North America. These events led the National Farmers' Union to charge in 1968 that when European companies were bought out by American capital, supplies to CCIL were cut off. With a secure agreement with Volvo for tractors and combines, CCIL made a similar point the same year when it noted that "we will never be caught out by American capital again." [54]

Sale of this imported European equipment forced CCIL to face still another aspect of farm machinery culture. These "foreign" machines, a revealing description that showed just how the U.S. and Canadian farm equipment markets were integrated, were not overly popular. Farmers worried about availability of parts since CCIL was forced to shift its brands. And, while CCIL continued to manufacture equipment, some observers after the mid 1960s charged—although not entirely fairly—that it was no longer the technological innovator that it had been in the 1940s.[55]

By the mid 1960s, the scope of the changes in CCIL's approach was becoming evident. In order to manage its network of dealers, in 1965 CCIL began to employ "district managers," its equivalent of the full line practice of using blockmen. It noted in 1967 that "we cannot afford to let up in the slightest degree the pressure to get out and sell," and by 1968, the company had a good distribution system staffed with salaried staff.[56] This shift in operating philosophy intensified after 1969 with a change in management, an event that revealed deep and long-standing personality conflicts within the organization.[57] The new management used extensive advertising, price cuts, and American management and marketing theory that emphasized personnel development and promotion as the means by which the company could break into the market in a significant way.[58] The new management was reluctant to abandon the traditions of the company, but they recognized that sales had to be made to keep the company alive, and this seemed to mean becoming more like the full line companies in style and approach.[59] Again, CCIL was being drawn into an approach to marketing equipment that it could ill afford. The company changed its name to Co-op Implements to give it a new look, and by 1980, it was giving cash rebates to purchasers and interest-free loans on some lines; the tactics full line companies had customarily used in depressed markets.[60]

While the company had traditionally been run as a tight ship, in the 1970s it became highly bureaucratic, leading to increased overhead. It also began, unlike its earlier practices, to rely on money borrowed from co-ops and governments to meet its problems. The early and mid 1970s were boom years for the farm implement industry and CCIL built a new plant in Winnipeg while continuing to import European tractors and combines.[61] The market collapse of the late 1970s forced the Prairie Provinces and the major prairie co-ops to set up a financial package to bail out CCIL in 1977. High interest rates in the early 1980s further restricted the company's recovery, and additional capital was injected into the company by the federal and provincial governments and the co-ops. By this time, CCIL faced a complex set of problems. It could not obtain sufficient equity capital, its manufacturing plant was operating at less than capacity, it was producing too many lines of equipment given its market share, and, among other difficulties, it was overstaffed. In 1984, the Winnipeg plant was closed and CCIL's manufacturing and engineering rights were purchased by Vicon Manufacturing Inc., a Dutch multinational. Vicon created a division called Cereal Implements at Portage la Prairie to manufacture and market CCIL equipment. CCIL (Co-op Implements as it was now called) maintained its depots, but gradually reduced their number to save costs. These tactics proved to be only of temporary benefit, and the company went into receivership in 1991.[62]

CCIL had originated because of prairie farmers' criticism of the way full line companies sold farm equipment. Although other factors in CCIL's ultimate failure cannot be downplayed, nonparticipation by farmers was crucial. It is an ironic paradox that, while farmers complained about the full line companies, when an alternative was presented, they were shown to be unwilling to abandon their traditional relationship with the industry and the distribution and sales system it had used for so long. This paradox, however, perhaps helps to explain the enthusiasm with which farmers embraced regulation of the industry through provincial farm machinery legislation. Farmers wanted the equipment the full line system provided—they wanted its style and its integrated system of distribution, sales, and repairs. Perhaps they felt that state regulation could minimize many of the problems they experienced with this system but would not change its essential elements. In this way, they only needed to make a few basic changes in the way that they purchased equipment.

An Early Prairie Industry

There were many efforts to manufacture farm equipment on the Prairies before World War II, but the overall success of the industry was elusive. A range of forces was responsible for growth after the war, including the creation of industrial infrastructure and changes in the types of equipment prairie farmers demanded. The prairie industry was made up mostly of short line manufacturers, who can simply be described as producing a limited line of equipment. Most served specialized, often uniquely local and regional needs that were rarely met by full line or other nonregional firms.

At the broadest level, short line manufacturers operated within the framework created by full line Canadian and American companies based in Ontario and the United States, companies that had integrated sales, distribution, and credit systems. Central Canadian manufacturers also at times received freight rate advantages in shipping equipment to the Prairies, and until 1944, they also operated behind protective tariffs. It was these factors that helped determine the level of industrial activity on the Prairies. Of equal, if not greater, importance before World War II was the limited industrial infrastructure and the orientation of the region's economy to primary production rather than manufacturing. Circumstance also played an important role because economic cycles affected the development and fortunes of regional manufacturing.

Farm Implement Manufacturing on the Prairies

Full line manufacturing was highly complex because of the number of components required to produce a wide variety of equipment. In this sense, the task of short line manufacturers was easier. Even so, some types of farm equipment were more complex and required a greater degree of precision and greater ancillary support industries than did others. Broadly, tillage equipment was the least complex to manufacture, while haying machinery, swathers, and feed-processing equipment were more complex. The greatest technical precision and most extensive range of support industries were required for equipment with diesel or gasoline motors and transmissions. The most popular short line activity on the Prairies has been in tillage equipment, haying machinery, swathers, and feed-processing equipment, while the manufacture of types of equipment with engines and transmissions has been less common.

Determining the level of prairie manufacturing of farm equipment before

World War II is difficult as activity tended to be short-lived and was often combined with other manufacturing. The *Census of Canada* reported only two firms in 1901 and six firms in 1911 that exclusively manufactured farm equipment. All were located in Manitoba. Beginning in 1918, the Dominion Bureau of Statistics (DBS) also reported the number of firms in each province manufacturing agricultural implements, but again only listed those manufacturing farm equipment as their primary product. These figures indicate only limited activity on the Prairies. In 1918, there was a total of about $74 million invested in the agricultural implement industry in Canada, of which about $1.2 million was in Manitoba and $44,000 in Saskatchewan and Alberta.[1]

Yet, almost from the beginning of European settlement on the Prairies, there was considerably more activity in manufacturing farm equipment than was officially reported. Much of it was informal and short-lived, making an objective assessment of its economic role difficult. However, a nonscientific tally of prairie firms advertising equipment they manufactured in various publications including *The Nor'-West Farmer, Canadian Farm Implements,* and *The Grain Growers' Guide* suggests considerable activity.[2] Between 1880 and 1900, 14 firms advertised implements of their own manufacture, and all except 1 were located in Manitoba. From 1901 to 1918, 120 additional firms were counted, and a further 71 advertised between 1919 and 1945. Although these firms were located throughout the Prairies after 1901, the greatest number were in Manitoba, especially Winnipeg. Some of these manufacturers were blacksmiths, foundries, and machine shops. Others were farmers working from their farm or a nearby town, and yet others were short line firms selling equipment that they manufactured. In many cases, locally manufactured products were sold along with equipment made elsewhere. While it would be wrong to see most of these firms as the start of an industrial base or as possessing much continuity, they reveal a much higher level of activity in invention and fabrication of farm equipment on the Prairies than do the official statistics.

The most important among these manufacturers were blacksmiths, small iron works, and foundries. These businesses produced a limited range of simple, small equipment such as land packers, harrows, plows, and water pumps. This was not a phenomenon unique to the Prairies: many of the implement companies established in the mid-nineteenth century in eastern Canada had grown out of blacksmith shops or foundries. Of these, Daniel Massey, who began manufacturing tillage equipment in 1847, was the most famous.

Blacksmith and metal-working shops were among the first businesses established in the prairie region after 1870. The Vulcan Iron Works in Winnipeg, for example, was established in 1883, while the Lethbridge Iron Works was set up in 1898. In 1905, there were 18 blacksmith/foundries in Manitoba alone, and by 1927, their number had increased to 22 firms in Manitoba and 15 in Alberta, mainly producing castings and sheet metal products.[3] By this point, these businesses were well established and a familiar part of most urban places on the Prairies. Their manufacturing activities were highly varied and localized, and in 1920, it was observed that "in most cases their output is determined by the requirements of the

Blacksmiths, such as this one in Virden, Manitoba, in about 1890, were an early feature of most towns. They repaired equipment and in many cases also manufactured it. (Virden - Business 2–1, PAM)

district they serve and in many cases they manufacture machinery which is patented by a local farmer for use in the immediate district."[4]

Broadly, there were three levels of metal work shops. At the top were well-equipped shops in cities and the largest towns such as the Vulcan Iron Works in Winnipeg and the John East Iron Works in Saskatoon. They performed relatively sophisticated metal work, including stamping out patterns, casting metal parts, lathe work, and custom bending. By the 1920s, some would also have offered

An illustration of the John East Iron Works machine shop in Saskatoon in 1937. (John East Iron Works Machine Shop Pamphlet B500, M7198, file 7, GAI)

welding services for repairing equipment. These same urban centres would also be home to smaller shops capable of a more limited range of work. Villages and small towns were the site of the least sophisticated shops; these shops did the most limited range of repairs and metal working.[5] Many of these metal works manufactured a few farm implements as well as providing a range of other services and so cannot be described precisely as a farm equipment industry. They were a mixed business type, combining fabrication and custom service.

More specialized, and fewer in number, were short line farm implement companies that exclusively manufactured farm equipment. A number were active in the early 1880s in Manitoba, although few were as successful as the large foundries and metal shops. The first short line manufacturer incorporated in the region was the Emerson Agricultural Works of Emerson, Manitoba. Organized in 1882 to manufacture harvesting equipment, the company appears not to have gone much beyond the incorporation stage.[6] A handful of other short line companies were established before 1900 in Manitoba. Among the successful ones was the Brandon Pump and Windmill Works. The company was established in 1893 by H. Cater, and following his death in 1945, his daughter became president of the firm and his sons vice-president and secretary.[7]

As with many early short line firms, the Western Manufacturing Co. of Indian Head, Assiniboia [Saskatchewan], was poorly capitalized and short-lived. It produced grain cleaning and treating equipment in 1904 and combined retailing with manufacturing. (*The Farmers Advocate,* July 13, 1904)

There was a surge in farm equipment manufacturing during the boom years before World War I and during the war when agricultural prices were high. DBS reported seven firms in Manitoba and four each in Alberta and Saskatchewan in 1918, the highest number it reported until after World War II. Our tally based on advertisements also shows a high level of activity, with 96 firms advertising farm equipment during the years 1908 to 1918. There also was a short burst of activity in the mid 1920s, but the Depression of the 1930s was a difficult period. DBS reported no farm equipment firms in Saskatchewan between 1932 and 1942. None were listed for Alberta between 1932 and 1935 and only one was listed thereafter until 1942. In Manitoba, the number of firms dropped from five in 1930 to two in 1933, followed by a staggered increase to four in 1942. Our tally reveals much the same pattern, although as would be expected, there was more activity than officially reported. Nevertheless, in the whole region, only 19 firms advertised prairie-made farm equipment between 1932 and 1942. While regulations during World War II restricted manufacturing of nonmilitary products, there was an increase in the number of farm implement firms reported by DBS after 1943. By 1945, it reported six firms in Manitoba, one in Saskatchewan, and three in Alberta.[8]

The Prairie Short Line Industry to the 1940s

The overall lack of continuity, success, or expansion of small scale manufacturing on the Prairies before 1945 was due to a range of factors. In the first place, the amount of equipment used on prairie farms before World War II was very limited in comparison to what became typical after the war. Statistics showing the amount of machinery on prairie farms are available only from the 1920s. By the early 1930s, the most common major equipment found on prairie farms were binders, distantly followed by tractors.[9] This sort of machinery was produced exclusively by full line companies. There was, of course, other equipment in common use: harrows, cultivators and plows, fanning mills, grain picklers for treating seed grain for smut, and grain crushers, as well as farmyard infrastructure like tanks, troughs, wagons, pumps, and windmills. It was this smaller equipment that early short line manufacturers provided.

For them, as for short line manufacturers after World War II, serving local needs was a central concern. As early as 1882, it was remarked in Winnipeg that the special equipment needs of prairie farming presented an opportunity for farm equipment manufacturing in Winnipeg. They "would have neither freights nor duty to contend with" and would be centrally located in the prairie market. They also had an advantage in being able to make "practical personal observation of the real wants of the county."[10]

Yet, the realization of this advantage was slow to emerge. Much of the small farmyard and field equipment produced by these early firms was little different than that produced elsewhere in Canada. By the end of World War I, however, it became clear that the equipment that was being manufactured by the full line firms or the nonprairie short line companies was not what was needed for tillage in dry land areas. The environmental disaster that befell the dry land areas of the Prairies in the interwar years was largely the result of tillage practices and their accompanying equipment. This failure had an important impact on prairie farm implement manufacturing and confirmed the need for a regional response. Tillage equipment was traditionally an important part of prairie equipment manufacturing, but the drought created a situation in which short line manufacturing expanded. While soil degradation caused by wrong tillage practices was not unique in North America to the Canadian plains, it was sufficiently restricted that local solutions, and unique equipment, were required.

Throughout the 1930s and early 1940s, there was considerable activity in developing and manufacturing tillage equipment suited to prairie conditions. A vital lesson had been learned: not all farm equipment technology had universal application, and the transfer of some of it in an unmodified form had drastic effects. But full line companies rarely undertook this work; the market for most dry land equipment was too small and specialized to be attractive. Instead, their concentration on equipment for general purposes created a gap that was filled by local inventors and manufacturers. The emphasis on tillage equipment that would become such a marked feature of prairie short line manufacturing after 1945 was the result of distinctive local needs not being met by the full line industry. In this sense, the isolation of the Prairies and its unique requirements assisted the development of farm equipment manufacturing in the region.

In fulfilling needs that other manufacturers neglected, a niche was opened for prairie manufacturers. Opportunity also came in the area of improving the quality of equipment in general. While in California in 1936, Charles Noble observed that

> I have been surprised to see how much of the farm equipment is designed here on the coast by the farmers and built in comparatively small shops. It is so much better than the frail designed tools made in the East to sell at competitive prices in a large volume. I am, and have been sure for some time, that we must do the same in two or three lines. Our little shops like the Riverside Iron Works [Calgary] could do a large business making simply constructed machinery with strength and capacity.[11]

Such hopes were widespread, and Noble was typical of prairie short line manufacturers and inventors in yet another respect. He was a farmer, as were most early inventors of prairie short line equipment. As a group, farmers showed a strong inventive and mechanical bent, and for a great number of them, the natural joy in the exercise of imagination found expression in working with machinery. Since repair parts were expensive or sometimes difficult to obtain, many were forced to adapt materials at hand or devise home-made solutions to keep equipment running. These farmer inventions and adaptations were widely recognized as holding important clues to improved agricultural technology.

Although some experimental farms designed equipment for seeding and harvesting the small plots used in scientific studies, their most important efforts were in testing and improving home-made machinery and assisting in improving their design.[12] In 1936, the Agricultural Engineering Committee noted that "the majority of progressive ideas for machinery improvements has come from farmers." Since there were over 200,000 farmers on the Prairies, "at least one percent of whom will have one or more useful and progressive ideas," the committee recognized the value of collecting and promoting these ideas.[13]

In most cases, these ideas modified existing equipment to overcome design problems or adapted it for prairie conditions. There were a great number of patents issued in connection with farm equipment. Saskatchewan residents alone patented a total of 3,200 inventions from 1905 until 1980. The peak period for this patent registration in Saskatchewan was the 1920s, when almost one-third of all patents issued before 1980 were granted. Most patents related to agricultural equipment, and the majority were designs of unique elements that improved existing equipment. Patents for new equipment were not common, perhaps because the general principles of a design were often not able to be patented.[14]

Farmers were the source of many inventions which sometimes became manufactured products. Others, such as this device for operating a binder in wet fields, were used only by the inventor. Photographed at Delia, Alberta, about 1916. (NA–2015–1, GAI)

Even though Canadian patent fees were the lowest in the European world, many farmers did not seek patent protection for their inventions.[15] There was apparently a widespread belief among farmers before World War II that patents mainly served to keep implement prices high by preventing competition.[16] Securing and defending a patent were also expensive and complex undertakings. The Hanson brothers, who farmed near Regina and developed the swather in the mid 1920s, decided not to patent their invention unless it was "something that could be granted a broad or basic patent right, something that could be defended in court if need be." As they reasoned, "a specific patent is not of much value unless you fabricate and sell and get returns that way. You may be sure that someone will snipe you from any angle possible. You may have to go to court to defend yourself. With this in mind we welcomed engineers from any of the implement companies, they were free to copy any or everything." As a result, their invention was picked up by IHC. Twenty years later, in a comment that revealed much of the practicality and caution that characterized prairie farmers of the interwar years, Helmer Hanson noted, "we still have our farms. We did not lose them in court trying to defend specific patents. It is easy to overestimate an idea that is so slow in coming that it looks too big when it does come." [17]

There was another problem with obtaining a patent that had much wider implications for manufacturing. The early farmer-inventors had plentiful ideas and were able to produce working models, but they often could not engineer the invention in the form required for patenting and production. Helmer Hanson noted that

> all our blueprints for the whole thing were stored between our ears, no drawings, each of us could see in his mind the complete machine and swather ready to go to work. Not being engineers or draftsmen, we did it the only way left to an untrained mind . . . We were handicapped by limited mathematics, true, we worked out our own problems but we probably did it the hard way—sometimes by sheer reasoning.[18]

This lack of formal engineering skill was a general characteristic of prairie inventors. The first agricultural engineering department in western Canada was set up at the University of Manitoba in 1906, and similar departments were established at the University of Saskatchewan in 1909 and the University of Alberta in 1921. Saskatchewan's department was the largest, but all provided training through degree programs and short courses for farmers. The discipline was divided into four fields, one of which dealt with power and machinery, while the other three emphasized farm structures, electrification, and soil and water conservation.[19] These programs were somewhat limited by financial restraints, and in 1939, Evan Hardy, an agricultural engineer at the University of Saskatchewan, contended that a lack of "competent engineers" was one barrier to the development of prairie farm equipment manufacturing.[20]

Engineering problems were compounded because a great deal of farmer invention before World War II was carried out under crude conditions. While almost all prairie cities and towns had installed electricity before World War I or during the interwar years, farms did not generally have electricity until after World War II. In the late 1920s, most of the farmers who invented machinery had few good tools, welders were unavailable, and acetylene torches were "for the big shops, not for farmers." A forge was used, only cast iron was available, and parts were fabricated or adapted from automobiles or other equipment.[21]

While this limited supply of infrastructure, tools, and raw materials was an inconvenience for farmers, it was a significant barrier for farm equipment manufacturing. In 1893, it was rumoured that the McCormick Co., Chicago, was planning a factory in Canada. Winnipeg's city council urged the company to select Winnipeg as the site, but McCormick replied that it was unsuitable because it was not central for the company's market and because "the raw material of which our machines are composed is largely steel and malleable castings" which were unavailable there.[22] This material had to be imported into the region, and in 1913, for example, Alberta Steel Products Ltd. of Dunmore (near Medicine Hat) manufactured metal grain bins using steel imported from Pittsburg.[23] In the same year, however, a steel mill was established at Selkirk, Manitoba, which was the beginning of the infrastructure needed for manufacturing farm equipment in the area. The plant was apparently unable to meet all of the varied needs of even simple fabrication, however, and manufacturing with metal remained difficult and expensive since most raw material had to be imported from a considerable distance.

By 1920, prairie blacksmiths and foundries were still importing most of their materials.[24] In the late 1930s, only cast metal was available in Alberta and steel was imported. As the Automatic Harvester Co., Calgary, which manufactured a header-barge (an implement for cutting crops for harvesting), told the public in 1937, "please keep in mind that a large portion of the material used in this machine is imported from a distance of 2,000 miles, necessitating time to deliver, and orders for materials have to be placed months ahead of the harvesting season." [25] Special materials also often had to be purchased in large volume. In 1937, the Lethbridge Iron Works attempted to obtain special-sized angle iron for C.S. Noble, but an order required a minimum of two tons and would take three to six months to receive. Nor could prices be guaranteed. In the same year, the Riverside Iron Works in Calgary informed Noble that material he wanted would have to be hand forged, and the cost, "we are afraid, would be prohibitive." Steel mills required large orders of at least a car load for specialized material since a special run had to be scheduled to produce it.[26]

The limited availability of energy for manufacturing also posed a problem for manufacturers. Even small scale manufacturing with iron and steel required a good supply of electricity. Manitoba was perhaps most advanced in this respect. Winnipeg had an adequate electrical supply for heavy manufacturing by about 1911, although the city's system was nearing capacity by the mid 1920s. The Manitoba Power Commission was established in 1919 and provided power to cities, towns, and villages, although it seems that capacity in most small towns was limited.[27] While most other prairie cities also had adequate power, small town electrical supply was usually insufficient for industrial use and many villages had no electrical service at all. Along with problems of supply of material, this was another constraint for industrial development. In 1939, C.H. Richardson of the Richardson Road Machinery Co., Saskatoon, a manufacturer of steel culverts and road-building equipment, told a committee of the Saskatchewan legislature that farm implement manufacturing in Saskatchewan faced "some unfavorable cost conditions as compared with the East." Alone, higher power costs were probably not significant, but combined with other factors they created major impediments. Labour cost slightly more, and "coke costs would be unfavorable, power unfavorable, and freight on raw materials would be greater than on the finished products being shipped in from the East." [28]

These limitations operated against a backdrop of financial limitations—a general unavailability of capital and little access to financing. In 1939, Evan Hardy argued that the difficulty of securing capital to establish a plant restricted the development of manufacturing in Saskatchewan. While encouraging industry was a popular objective for local governments before 1945, this usually involved offers of low cost land and tax relief but rarely cash or loans for start-up costs. Brandon's grant of inexpensive city land and tax concessions to the Brandon Machine Works for the establishing of its plant there in 1901 was a typical example of this process.[29]

Senior governments rarely offered financial assistance for prairie manufacturing before World War II. The federal government, however, played an indirect role in assisting industry by subsidizing infrastructure like railways and through policies such as protective tariffs. It also encouraged research to increase agricultural production in the region. In the late 1880s, it set up experimental stations at Brandon and Indian Head, and numerous others were established later. Through plant and animal research and other programs, these stations made a significant contribution to knowledge and the development of the economic infrastructure of prairie agriculture.[30]

Provincial governments also followed similar approaches, but both levels of government generally provided little direct encouragement to small industry. A significant change in policy came in 1925 with the formation of the Manitoba Industrial Development Board. This provincial agency surveyed industrial opportunities and encouraged investment in manufacturing in the province, but its effectiveness seems to have been stymied by the Depression.[31] Elsewhere on the Prairies, the creation of these provincial agencies was slower. It was not until the late 1930s that the government of Alberta became involved with such efforts, largely in an attempt to apply Social Credit economic theory. In Saskatchewan, a bill for establishing an industrial development board to encourage industrial development was defeated in the Assembly in 1939.[32] Private agencies like the CPR also actively promoted economic development, although their programs were tied solely to corporate objectives.

Clearly, manufacturers and inventors before World War II could not look to senior governments for much financial assistance in establishing a business. Capital had to be raised privately. Small short line prairie firms seem rarely to have been able to borrow capital from banks and other financial institutions, and a potential manufacturer had to raise money elsewhere. *The Nor'-West Farmer* noted in 1919 that while inventors "have not an easy time as a rule getting people with money to back their invention . . . the thing can be done if one goes about it persistently." [33] It was common for inventors to try to interest urban investors and farmers; the farm press often contained advertisements by manufacturers asking for investment capital. Sometimes these advertisements argued that the small manufacturer, like the farmer, was pitted against the full line companies and that the development of regional manufacturing would help reduce prices. The Canadian Farm Implement Co., Medicine Hat, used this approach in 1920. Advertising itself as "Not in the Trust," it urged farmers to buy its shares and "help yourself by becoming a partner in this Western Industry. Control the price of farm implements, control the profits, and share in those profits . . . We want every farmer to join in our movement to break the price of farm implements." Shares cost $1 each, sold in lots of five shares.[34]

While the level of investment that this offer garnered is unknown, the company was short-lived. It seems, however, that farmers were willing to invest in implement firms when they had spare cash, even though it was a risky investment given the nature of the industry at the time. There were occasional complaints by farmers about losses on investments, but details are sketchy.[35] The Seed Grain Grader Co., incorporated in Regina in 1915 to manufacture a grain grader, demonstrates the fate of one company with farmer investment. Of the firm's issued shares, 220 were held by the inventor, C.A. Torrance. Farmers, mostly from the Wolseley area, invested in one-half of the remaining shares, and the balance was held by Regina lawyers, "managers," and "agents," probably as payment for selling shares. This reveals the low cash investment in the firm. The shares held by the inventor were issued in lieu of cash for purchase of his patent, and the shares issued to the "agents" did not bring in any cash. As a result, of the company's stated capital account of $1,500 at incorporation, only about $800 represented a cash investment. And, since a sizeable number of shares were paid up to only half of their value, much of the firm's capital was in promissory notes. The company folded soon after its incorporation.[36]

Because of problems with capitalization, many people handed over fabrication of their inventions to a local foundry, blacksmith, or sheet metal shop. This lack of working capital also often led to distribution of the product in the same manner. Inventors usually could not tap into existing dealerships for distribution. Under threat of losing their contract with their full line supplier, implement dealers often did not handle any other company's equipment.[37] In the trade, this was called "full line forcing." It restricted competition and helped eliminate smaller manufacturers because they could not sell their products easily or provide customers with servicing and repairs.[38] This negatively affected all short line manufacturers alike, regardless of their location in the country. Because of this, regional foundries or blacksmiths often took on the job of distributing—on a royalty basis—the products they manufactured under license from the inventor. This was not, however, always a satisfactory approach. For example, the Riverside Iron Works manufactured and marketed a land packer on this basis for C.S. Noble. But, others were also manufacturing the same packer, which led Riverside to give Noble an ultimatum in 1937. If he did not defend his patent and stop others from manufacturing his packer, Riverside would stop paying royalties since this put them at a disadvantage in selling it.[39]

Small manufacturers had to find other means of distribution. Some sold equipment through advertisements in the farm press, often combining these direct sales efforts with requests for agents and retailers to handle their products. Large fairs and exhibitions also were a regular opportunity to display and demonstrate products. There was usually an "industrial" section featuring farm equipment, and local manufacturers took advantage of these venues.[40] For instance, Manitoba Engines Ltd., which in 1908 was manufacturing grain grinders and other products, had a permanent exhibit at the Brandon exhibition and also showed at the Winnipeg exhibition.[41] In the interwar years, these advertising approaches were supplemented by other techniques. For example, in 1938 the Morris Rod Weeder Co., a small Saskatchewan firm manufacturing a rod weeder with unique features, had salesmen on the road who drove "through the territory, taking with them a Morris rod weeder arranged on special air tire road wheels." It was shown to dealers, and field demonstrations were arranged for the community.[42]

These approaches provided something of a solution to the problems of distribution faced by short line manufacturers. Other problems proved more difficult. Farmers usually could obtain credit to purchase machinery only from the manufacturer, but poorly capitalized short line companies could not afford to offer any financing. Most sold only for cash, and often manufactured on a pre-sold basis, with cancellation permitted because of crop failure.[43] Most full line companies, however, not only offered credit but often gave no discount for cash. This use of credit to attract customers made cash sales, as needed by short line firms, less attractive. This problem was not unique to prairie short line manufacturers given full line company domination throughout the country. The Sylvester Manufacturing Co. of Lindsay, Ontario, which manufactured a mobile threshing machine around 1908, experienced financial trouble because of this "common practice of giving extended credit, running from two and three years, in selling implements." This liberal use of credit was offered "in lieu of lower prices." Since foreign sales tended to be for cash, the company attempted to gain export markets.[44]

Many newly established short line manufacturers also had to work hard to convince purchasers that they would be able to provide ongoing services. One of the greatest advantages possessed by the full line companies was their permanence and continuity. Given farmers' concerns about future supply of repair parts, this no doubt favoured full line manufacturers when it came to competing for trade. Since short line products were rarely in direct competition with those of the full line companies, this problem was of greatest significance when short liners tried to sell either equipment to retailers or component parts to other manufacturers. When the UGG was selling farm equip-

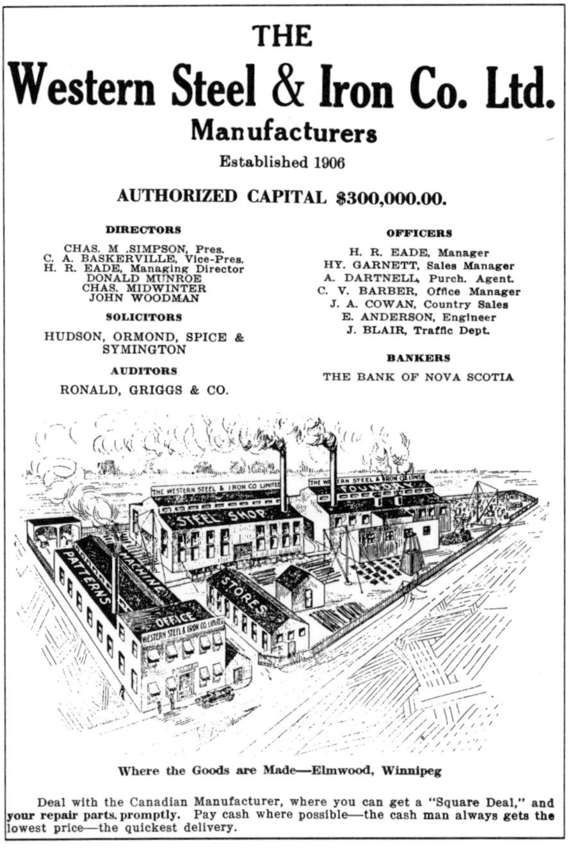

Early short line manufacturers sometimes used the same symbolism as the full line companies to project an image of success and permanence. (The Western Steel & Iron Co. Ltd., brochure, Ag.12.II.B.92, SABS)

ment, it tried to favour Canadian goods, but it was reluctant to deal with small short line companies. Aside from the full line firms, there were only a very few short line companies that it bought from since it considered it unwise

> to secure machinery from any plants that are not established on what may reasonably be considered a secure and permanent basis, as the difficulty of securing repairs in case they should go out of business has already been experienced. Further, frequent changing of lines is extremely unsatisfactory, necessitating as it does the carrying of an increased stock of repairs.[45]

In combination with other forces, these conditions limited the expansion and security of prairie manufacturers.

Prairie farm equipment manufacturing before 1945 developed within a craft tradition of organization and fabrication. Isolated from each other, numerous individuals invented equipment and often manufactured it on a small scale to meet local needs. Most, however, were not able to penetrate the market dominated by large, highly capitalized, and well-organized full line firms. There were a few efforts to assemble equipment, notably tractors, before 1920, but such approaches were rarely successful. Most complex farm equipment tended to be made by a single corporation that manufactured all the component parts needed. Even so, such heavy manufacturing required linkages to related sectors for a supply of raw materials and energy. For short line manufacturers on the Prairies, these were relatively limited before World War II. As well, prairie farm equipment manufacturers were hemmed in by the distributing and servicing system used by the full line firms. All of these factors contributed to limit prairie developments to the scale and methods of a craft tradition. As a result, although some individuals made important contributions to technological development and invention of prairie agricultural equipment, they did not form an industry possessing continuity or mutual dependence.

Growth and Development: Prairie Short Line Manufacturing After World War II

The rapid growth in farm equipment manufacturing on the Prairies after World War II, and especially after 1961, resulted from an increasingly widespread infrastructure for manufacturing along with a growing maturity of the prairie economy. A significant shift in the nature of farm equipment used in Canada, and on the Prairies in particular, was also an important force. While a handful of companies such as Morris Rod Weeder Co. and C.S. Noble bridged the two periods, most successful companies after World War II were newly formed.

The period after 1945 saw the beginning of sustained growth of short line manufacturing on the Prairies. In 1951, there were a total of 81 farm equipment plants in Canada, 36 of them on the Prairies. Yet the prairie industry was still dwarfed by that of Ontario, which had 94 percent of the value of production.[1] While the overall number of prairie firms decreased between 1951 and 1961, the proportion of farm implements manufactured in western Canada of total Canadian production increased from 5 percent in 1952 to 14 percent in 1963. In Manitoba, growth in this sector was higher than in other industrial sectors in the province. Winnipeg, the centre of the industry, had a head start in distribution and manufacturing that gave it an advantage in an expanding market, and it remained the most central location for the wheat-growing areas of Canada and the United States. It had a substantial labour force and wage rates were relatively low. Freight costs were also generally lower between Winnipeg and most regional market centres than from points further east. The value of production doubled between 1961 and 1971, but the most significant period of growth was from 1971 to 1981 when the value of production increased 10 fold. As set out in Table 5–1, after 1961, Manitoba was second only to Ontario in value of factory shipments of farm equipment, and the gap between the two provinces narrowed significantly. In 1958, Manitoba had produced 5 percent of Canadian factory shipments of farm equipment, but by 1981, it was producing about 28 percent, and economists were beginning to call the farm equipment industry a key sector of Manitoba's economy. In 1979, Versatile and CCIL, the two largest farm equipment manufacturers in Winnipeg, had 1,305 and 874 factory employees, respectively.[2]

TABLE 5–1
Value* of Production and Number of Implement Plants Canada and the Prairies 1961–1981

	1961		1971		1981	
	Number of firms	**Value of shipments**	**Number of firms**	**Value of shipments**	**Number of firms**	**Value of shipments**
Manitoba	15	$9m	24	$37m	25	$395m
Saskatchewan	5	$1m	23	$12m	30	$158m
Alberta	7	$2m**	18	$7m	22	$57m
Ontario	27	$123m	45	$174m	77	$733m
Other Provinces	13	$3m	20	$10m	53	$59m
Total Canada	67	$138m	130	$240m	207	$1,402m
Total Prairies	27	$12m**	65	$56m	77	$610m
Prairies % of total	40%	9%	50%	23%	37%	+ 44%

Source: Compiled from Statistics Canada, *Agricultural Implement Industry. Annual Census of Manufacturers* for years cited.
Note: All figures do not total because of rounding. These statistics reflect firms manufacturing only farm implements and therefore show less production than was actually taking place.
*Value of shipments of goods of own manufacture.
**Combined with British Columbia.

In 1971, the number of firms on the Prairies had more than doubled over the number in 1961, and by 1981, their total equalled the number operating in Ontario. More significantly, the value of production on the Prairies was reaching that of Ontario. Manitoba was still most important in prairie production, where several firms had become major players in the Canadian farm equipment industry. But it was increasingly challenged by Saskatchewan where farm equipment manufacturing was rising steadily, especially between 1971 and 1981. In 1971, one survey revealed 34 companies manufacturing farm equipment in Saskatchewan, employing 1,400, and having $14 to $16 million in annual sales.[3] This is considerably higher than the number stated in Table 5–1, which includes firms manufacturing only farm equipment.

Just as before World War II, none of these prairie manufacturers were full line companies, which remained firmly based in Ontario. Nor did full line companies see advantages in moving their production west. One historian of the industry has noted that "no significant effort was ever made by a leading Canadian implement company to remove any of its manufacturing operations to Western Canada." [4] The

In 1965, the Morris Rod Weeder Co. Ltd., Yorkton, was a successful and expanding manufacturer of tillage equipment. This photo shows the steel warehouse at the Yorkton plant. (*Saskatchewan Diamond Jubilee Directory,* 1965, p. 185)

reasons for this varied. Ontario was closer to raw materials like steel and to component-part manufacturers. In addition, the western Canadian labour force was small. As one informant with a full line firm told the province of Manitoba in 1967, if the prairie farm implement market was twice as big, and if Manitoba had a larger labour pool and an iron and steel industry, "they could justify a manufacturing operation in Manitoba."[5] These factors had long affected farm implement manufacturing on the Prairies, and for short line firms, some of the factors that had shaped the industry before World War II continued to be significant.

Postwar Manufacturers

The success of postwar short line manufacturers obscures some elements of continuity between the two periods. For one, the connection between blacksmiths and farm equipment manufacturing remained strong. John Schulte, who took over his father's blacksmith shop and implement dealership in Englefeld, Saskatchewan, in 1938, began manufacturing brush-cutting equipment in 1942. He eventually expanded into other lines, the most successful of which was rock-picking equipment, and in the 1980s, Schulte Industries began production of a rotary cutter. By 1972, the company had about 55 employees, and by 1987, sales were $2.5 million. This growth was in sharp contrast to most prewar prairie manufacturers, as was Schulte's substantial sale of goods into export markets.[6]

Another manufacturer to build on the blacksmith tradition was Leon Malinowski, who took over his father's shop in Bankend, Saskatchewan. In the early 1950s, he was manufacturing stock troughs, fuel tank stands, and similar products. By 1952, he was also manufacturing dozer blades of his own design that had a quick mounting system as well as special features to help prevent breakage if the blade hit a heavy object. He incorporated Leon's Manufacturing in 1965 and moved to

Yorkton in 1967. By this time, the company was manufacturing a range of tillage equipment in addition to dozer blades. By 1989, it had about 150 employees and substantial export markets.[7]

The farmer-inventor tradition also continued to be central to prairie manufacturing. It was a tradition undimmed by the other changes taking place in agriculture and rural life. In 1953, CCIL noted that every week it received suggestions from farmers with "ideas for new machines." [8] For those with a combination of technical knowledge and business acumen, these ideas often led to industrial application. Wilfred Degelman, for example, began manufacturing a rock picker on his Saskatchewan farm in 1963. Two years later, he set up a plant in Regina, and business quickly expanded. By 1973, Degelman Industries employed about 200 and exported about half of its production. Its sales had reached $9 million by 1975.[9]

Most postwar firms attempted to diversify into equipment serving both farm and nonfarm markets. Schulte Industries Ltd., Englefeld, Saskatchewan, developed this rotary mower in the mid 1980s to appeal to such a wider market. (Schulte Industries Ltd. Photograph)

These manufacturers demonstrated the persistence of earlier prairie traditions in yet another respect: they usually designed solely with an eye to local conditions. In 1962, Peter Anderson, the owner of a machine shop in Southey, Saskatchewan, explained the advantage of this approach. For him, inventing initially involved constructing a small working model, applying for a patent, and fabricating a full-size model. The most essential step, however, was to "test it in your own community among people you know." They would provide suggestions for improvement, and "if it is not workable or is not accepted by people in your own district, it will not likely sell further afield."[10]

While many of these firms carried on earlier traditions, their rapid growth and size were in sharp contrast to their predecessors from before World War II. As well, practices began to appear on the Prairies that owed little to the farmer-inventor or blacksmith traditions. For one, the realization by full line companies of the cost advantages of contracting out production expanded the opportunities of small specialized firms. In this system, one company manufactured equipment that was sold under the brand name of another. This had been done on a piecemeal basis by many prairie foundries before World War II, but none had manufactured exclusively for this purpose. Although declining sales volumes for the full line companies by the 1970s stimulated greater use of this approach, the practice had emerged almost immediately after World War II.[11]

One example of this new type of company was Killbery Industries Ltd., established in 1949 in Winnipeg. It manufactured sprayers, swathers, tillage equipment, and pickup reels for combines and swathers. Although about 65 percent of this production was sold on the Prairies, none of it appeared as a Killbery product. Rather, it was sold by full line companies under their brand names. There was substantial demand for such production, and in 1967, Killbery had sales of $4 million. However, falling markets led to financial difficulties and the company attempted to find other markets. This was difficult because its name was unknown to consumers (even though many of them would have been using Killbery manufactured equipment). In the early 1970s, the company attempted to diversify by going into snowmobile equipment and began promoting itself as a brand name in export markets. At this point, the company was purchased by Joe MacDonald, who renamed it MacDon Industries Ltd. It continued to manufacture primarily on contract to full line companies. With better farm equipment sales in North America in the 1970s, the company grew, and by 1976, it was producing all of Massey's swathers as well as some for IHC.[12]

Similarly, Agristeel, a company in Minnedosa, Manitoba, enjoyed substantial sales (estimated at $3 million in 1967) through contract manufacturing.[13] It manufactured tillage and seeding equipment sold by Cockshutt and Case, and it also manufactured some equipment for CCIL in the 1970s. Indeed, the range of prairie suppliers used by CCIL illustrates the interdependence and linkages that were emerging. From 1974 to 1976, CCIL continued to manufacture much of its own equipment, but also purchased packers and hydraulic drawbars from Inland Steel and Forgings in Winnipeg, harrows from Farm King at Morden, press drills and some combine components from Terra Implements at Minnedosa, and, among others, front-end loaders from Leon's Manufacturing at Yorkton.[14]

Most of these companies also manufactured their own brand name equipment, but the most successful prairie brand name manufacturer was Hydraulic Engineering Co., renamed Versatile in 1963. The company was established in Toronto by Peter Pakosh and Roy Robinson in the late 1940s. Pakosh had worked for Massey in its tool design department, but frustrated with the nature of the Massey design system, he applied his ideas elsewhere. By 1947, the basement of his Toronto house had been taken over by his passion for designing equipment, and in 1949, he and Roy Robinson set up a plant in Toronto to manufacture grain augers and other small equipment. Since most of the market was on the Prairies, the company moved to Winnipeg in 1952, which helped lower their freight costs and allowed for fast service on parts. In its first year in Winnipeg, the company had sales of about $50,000.

Initially, Versatile manufactured sprayers and grain augers but soon added harrow drawbars to its line. These were traditional prairie short line items, but the company came out with a self-propelled swather in 1954 and a pull-type combine in 1963. The mid 1960s were a decade of significant growth for Versatile. In 1966, with around 600 employees and annual sales of about $12 million, it sold 3,000 self-propelled and 2,000 pull-type swathers. That same year, it began producing tractors, which traditionally had not been manufactured in Canada. The company had, to all intents and purposes, become a full line firm with its products marketed under its own brand name. In 1976, when Pakosh and Robinson sold the firm to Cornat Industries of Vancouver, Versatile had two factories in Winnipeg producing swathers and tractors, and one in Fargo, North Dakota, producing axles and transmissions.[15] In the tough farm equipment market of the early 1980s, the company was not able to sustain its success, and it was eventually sold to Ford New Holland, a company created in 1986 with Ford's purchase of Sperry New Holland, a large short line manufacturer of haying equipment. The acquisition of Versatile gave Ford the North American tractor factory that it had so far lacked. And reflecting the continuing amalgamation and mergers that still characterized the full line industry, in 1990 Ford New Holland and FiatGeotech merged.[16]

In some respects, Versatile began as a typical prairie short line firm. It had a limited range of specialized products, and after its move to Winnipeg, it was close to its market. It was also rooted in prairie rural culture since Pakosh had grown up on a Saskatchewan farm. Versatile's success rested on a number of factors: the personality and drive of its owners; efficient operation with low administrative and selling costs; good design that appealed to farmers; its good fortune in being at the right place at the right time when general changes were taking place in both agriculture and manufacturing on the Prairies; and operation within a general period of buoyant farm equipment markets. Pakosh and Robinson were a well-matched team with a seemingly instinctive understanding of the prairie farm equipment market. Pakosh was very inventive and Robinson was an accomplished manager. And unlike the full line companies, Versatile was not reluctant to compete on price. Its self-propelled swather was the lowest priced on the market and in 1970 its combines sold for 20 percent less than competing models. Versatile also gave dealers a higher commission to encourage sales.[17] These tactics served Versatile well, preserving its sales volume even during a very poor sales year in 1969.[18]

The development of firms like Versatile and MacDon Industries demonstrated a changed environment for prairie farm equipment manufacturing. As before World War II, the majority of farm equipment in Canada continued to be sold on the

Prairies, but the postwar boom in mechanization of prairie farming brought important changes in farmers' expectations about machinery. With the lessons of the drought of the 1930s clearly in mind, postwar farmers demanded equipment designed specifically for prairie conditions and needs.[19] While demand for major equipment like tractors and combines declined with greater farm size, it increased for short line equipment like augers, front-end loaders, and some tillage equipment.[20] Indeed, George Morris observed that full line companies were disadvantaged in such a market. Short line manufacturers like himself had "identified a specific area and we specialize in that area." The full line companies were "so busy competing with each other trying to cover every market, there's no way they can bother to concentrate on these specific situations."[21] In addition, as Morris intimated, company size was important. As Wilfred Degelman observed, "a market segment that may not be attractive to the mainliners might be very attractive to a company of our size," which could respond relatively quickly.[22] The advantages of local and regional sales and specialized equipment also offset potential advantages of larger scale production, which were, in any case, less in farm equipment manufacturing than in some other industries. Economies of scale sometimes worked in reverse, especially in a market with declining sales volumes. A production run that might be profitable for a short line company might be too small for the large factories of the full line companies where an identical production run would be too costly.[23]

The Evolution of Infrastructure for Manufacturing

These developments in the postwar farm equipment industry took place within changes in the region's manufacturing infrastructure. In 1945, Winnipeg was the only prairie centre with significant industrial capacity. When CCIL was established in the mid 1940s, Saskatoon was designated as its head office, but the company was located in Winnipeg instead. Looking back, some observers in 1968 contended that the plant should indeed have been located in Saskatchewan, the heart of the territory the firm served. These criticisms failed, however, to recognize the changes that had taken place after the war. In the 1940s, a plant similar to that in Winnipeg was not available in Saskatchewan, and the raw materials and most of the components needed for manufacturing were produced in Winnipeg, Ontario, or the United States. At the time, this made Winnipeg the central prairie point for manufacturing.[24]

One crucial factor in heavy manufacturing was the availability of steel. While cast iron and sheet metal were produced at a number of places on the Prairies at the end of World War II, rolled steel was produced only at the Manitoba Rolling Mill at Selkirk, the only steel mill between the Soo and Vancouver. Using scrap as the basic raw material, it had an annual production in 1954 of 70,000 tons of steel in various forms. Although the Selkirk mill met much of the prairie demand for these products, it did not produce complete lines of steel.[25] The spread of steel-making facilities beyond the Winnipeg area in the 1960s was a notable development. In 1962, the Interprovincial Steel and Pipe Corporation (IPSCO) set up a mill at Regina. It used scrap metal, of which there was an adequate supply for prairie needs.

The plant had a significant impact on industrial development in Saskatchewan and improved Regina's advantages as a manufacturing point.[26] IPSCO was an important supplier of steel for local manufacturers; in 1973, Degelman Industries purchased about 60 percent of its metal from this source.[27]

Even so, shortages of iron and steel as well as other materials in the early 1970s created difficulties for prairie manufacturers. However, since these shortages were being experienced everywhere in Canada because of a booming economy, they did not represent a regional disadvantage. Degelman, for example, reported in 1974 that "every time we turn around we're out of something." The same problems were reported by CCIL, Versatile, and Morris Rod Weeder Co., where production was sometimes halted because of shortages.[28] These shortages did not last, and new materials soon were as significant as traditional ones like iron and steel. The increased use of rubber and especially plastics in manufacturing provided an alternative to traditional materials and greater ease of fabrication. A number of rubber, plastic, and fibreglass suppliers and fabricators emerged on the Prairies by the 1970s, often doing work for farm equipment manufacturers.

The increased availability of materials after World War II was significant, but it was especially important because of changing methods of fabrication which demanded less specialized mill work. Before World War II, most parts were cast from metal, were bent, turned on a lathe or a grinding wheel, or were made from wood. These parts were then bolted or riveted together. While these methods continued in use after the war, by the 1950s, they were giving way to welded fabrication using arc welders. Wood soon disappeared almost totally as a material of construction. Welded construction created lighter and stronger equipment and also required less extensive plant equipment and labour. Some parts could also be produced in-house with oxyacetylene flame cutting of metal.[29] Such manufacturing techniques were beginning to appear on the Prairies by 1940. In that year, the Morris rod weeders first began using welded frame construction. They were manufactured on contract by a firm in Manitoba.

These practices were not confined to factories. They came into general use because of the increased supply of electricity in both rural and urban areas. While slightly more farms in Manitoba had electricity than in Saskatchewan or Alberta, conditions in Alberta were typical. In 1941, only 5.5 percent of farms had electricity, mostly from individual generating plants. A few farmers had gasoline-powered generators or windchargers to generate electricity, but this power was expensive and was used largely for household conveniences and lighting.[30] Following World War II, the involvement of provincial governments and provincial power commissions led to a rapid increase in the connection of farms with line power systems. Manitoba's plans to extend electricity to farms were postponed because of World War II, but after the war, it aggressively extended electricity throughout the province. In 1947 alone, the Manitoba Power Commission connected 3,500 farms to electrical lines, and 50 times as many farms had electricity by 1949 as in 1942. A similar pace of development was also occurring in Saskatchewan and Alberta. On the farm, the first noticeable change was in lifestyle and home conveniences,

Welded steel fabrication became important in farm equipment manufacturing after World War II. A more flexible system, it produced sounder equipment than earlier systems that used rivets and bolts. Here a welder is manufacturing equipment at Farm King Allied Ltd., Morden, Manitoba, in the 1960s. (John Buhler Inc. Photograph)

but electricity soon dramatically lessened the physical drudgery of farm work and stimulated further farm mechanization. Because of electrical equipment like drills and welders, it also expanded a farmer's opportunity to fabricate and repair equipment.

Even before most farms were linked to electrical lines, many farmers wanted to have welders. In the 1940s, the University of Saskatchewan published a study on the potential farm use of generators for producing electricity. Using this idea, Clem Roles, associated with the university's engineering faculty, obtained (by accident and while on his honeymoon) several war-surplus generators. He recognized that with some modifications they would be suitable for powering welders. Roles adapted the generators so that they could be operated from a tractor belt, and he created an affordable welder outfit for use on farms without line electricity. Along with his partner, Tom Smith, Roles established a company, Smith-Roles Ltd., to market the welder outfit.[31]

The welder outfit was immediately popular and remained in production beyond 1960, although demand began to fall by 1954 because of rural electrification. Smith-Roles then began manufacturing welders that ran off line current. This was, however, more than just a welder. It was, Roles claimed, "a virtual machine shop in itself." It could be used to weld metal as well as solder, cut, braze, and heat it. Among other jobs, it could also charge batteries. With a five-year guarantee, it became one of the main products of the firm for many years. To go along with this equipment, Roles wrote a book (still in print) on arc welding for tradesmen and farmers. The company expanded by acquiring the Blanchard Foundry in Saskatoon in the early 1960s, as well as other companies, and it manufactured a range of field equipment and machine accessories until the late 1980s when it went into receivership. The firm continues under new ownership.[32]

The spread of electricity gave farmers the opportunity to work at a mechanical level rarely possible before. Modern machine tools and welders began appearing in many farm shops, increasing the technical level of innovation and the range of repairs that farmers could carry out. It also created opportunities for farmers to begin manufacturing farm equipment on their own farms. Other opportunities also arose with the greater flexibility and capacity offered by province-wide electrical systems. By the early 1950s, machine shops in the larger centres outside the cities were able to increase their range of service. At the same time, however, village shops began going out of business. Farmers were doing more repairs themselves and the larger and more sophisticated machine shops in the larger centres were now more accessible because of postwar highway improvement.[33] Yet, as early as 1947 in Manitoba, new small industries were appearing in smaller towns, a development credited to rural electrification.[34] Thus, while electricity represented one aspect of the technological change that was contributing to increased farm size and rural depopulation, it also created possibilities for the development of a more diversified rural economy.

The broadening of the industrial base in western Canada after World War II increased the possibility of success for short line firms. While these developments did not necessarily make raw materials less expensive on the Prairies than in Ontario, the proximity of short line companies to their market reduced order time, inventory, and therefore cost. Increased use of component parts after World War II confirmed these cost advantages. It was less expensive to bring in basic components than manufactured implements, and when other factors were equal, prairie manufacturers had a price advantage in the local market.[35]

The increased use of component parts was a significant change in prairie farm equipment manufacturing. Only a few short line firms before World War II had been able to use assembly systems, and it had never been a widespread or general practice. After World War II, a greater standardization of engine components emerged. Versatile's tractors, for example, used engines and gears manufactured by U.S. and Canadian companies, and it obtained its castings from foundries in Winnipeg. As a result, the company was able to combine assembly with fabrication to an extent unusual on the Prairies at earlier times and reduce the tooling costs for manufacturing complex equipment.[36] In addition to Versatile, other large firms such as CCIL and Morris used component parts manufactured outside their plants. There was, however, no general pattern in this use of components. When warranted by production or cost requirements, some firms began to manufacture parts that they had formerly purchased. Nonetheless, the substantial use of component parts led to high demand and prairie firms increasingly manufactured them. In 1973, for example, Leon and Ray Malinowski, the owners of Leon's Manufacturing, established Ram Industries in Yorkton to manufacture hydraulic cylinders for sale to other manufacturers.[37]

Like shortages of material during boom times, lack of skilled labour sometimes created bottlenecks in manufacturing. A skilled industrial labour pool on the Prairies was relatively limited outside of Winnipeg, and one manufacturer commented in 1967 that there had been "little improvement in the skilled labor situation in the west over the last 20 years."[38] This condition persisted, and in the early 1970s, it became worse because of the industrial boom everywhere in Canada. In 1974, Morris found skilled and even unskilled workers hard to keep or recruit. CCIL too found it difficult to attract skilled metal workers, and Versatile was short 50 or 60 workers that year.[39] Despite these problems, many companies found prairie workers to be hard working, loyal, and non-unionized. In farm communities, they were also available on a seasonal basis. These factors fit well with the cyclical qualities of the industry, and prairie manufacturers often cited them as the advantages of manufacturing on the Prairies.

Recurrent problems with availability of trained workers were addressed on a number of fronts. Government was able to assist directly, and Manitoba, for example, had loans and other programs for worker training by 1967. And in the 1970s, the newly formed Prairie Implement Manufacturers Association began working with government and educational institutions to ensure that training and educational programs met the industry's needs. Management skill was equally important, and some prairie firms encountered problems in this area as their operations expanded and became more complex. In 1968, it was noted that "management is, in many cases, a weak spot in the Manitoba farm equipment industry." This led to "poor planning, a poor distributive structure, inadequate financial control and a lack of real interest in growth and market development."[40] Increasingly, successful firms found it necessary to employ managers to handle administrative and financial matters. To assist firms in acquiring such management skills, governments developed management training programs to parallel other training programs.

Growth, Marketing, and Research

Government training programs were only one aspect of changed government policies on industrial development. After World War II, the Saskatchewan government became directly involved in promoting economic diversification. In 1946, it established the Economic and Advisory Planning Board to advise government on economic, industrial, and commercial matters and to formulate long-term planning for the province's economic development. Attracting established out-of-province farm equipment firms was not feasible, and it seems to have been recognized that home-grown development would be more stable and better adapted to the province's needs.[41] In this, Saskatchewan provided some assistance through its Industrial Development Fund, which made loans to provincial manufacturers to encourage economic diversification. Some farm equipment firms received assistance: Flexi-Coil Ltd., a Saskatoon manufacturer of a unique land packer, as well as Sakundiak Farm Equipment Co., Regina, received loans for plant development during the early 1950s.[42] In 1963, the Saskatchewan government confirmed these approaches through the creation of the Saskatchewan Economic and Development Corporation which made loans and grants for start-up capital for new businesses. It also extended funds to established industries for plant expansion and for research and development.[43]

Other provinces were also moving in the same direction. By 1967, Manitoba had set up programs to provide loans for marketing and product development and was aggressively trying to encourage the development of the farm equipment sector.[44] The federal government too had various programs, most notably under the Department of Regional and Economic Expansion (DREE) from 1969 to 1987. The DREE programs were complex, sometimes contradictory, and sometimes even detrimental since they encouraged overcapacity in some sectors. They nevertheless presented a co-ordinated approach to the government's regional development programs. Some prairie farm equipment firms received funds under these programs. Versatile, for example, received $4.6 million in 1978 to expand its Winnipeg plant.[45] There was, however, little agreement within the short line industry about these programs. Some saw them as essential, while others viewed them as providing unfair advantages to their competitors.

The applicability of these programs to farm equipment manufacturers suggested that financing in the industry remained a problem. Sales of farm equipment were seasonal (most equipment was sold in the first part of the year), which often meant a slow turnover of inventory. Some larger firms were able to generate sufficient revenue from sales to obtain conventional loans. And Versatile, alone among prairie farm equipment firms, issued stock to raise capital. The company went public in 1963. Its share offering was successful since Versatile's 9.7 percent return on assets during the late 1960s was highly attractive.[46] But for many companies, lack of capital remained a barrier. In 1967, Robin Nodwell Manufacturing Ltd., a relatively large Calgary farm equipment manufacturer, noted that financing was the greatest problem facing prairie short line manufacturers. Both start-up and operating capital were difficult to obtain. It argued that while U.S. short line manufacturers could usually borrow up to 90 percent of the value of their receivables, Canadian firms were fortunate to be able to borrow half that amount. This was especially significant since most dealers took short line equipment on consignment rather than buying it outright.[47]

The difficulties in finding capital were made worse by the chronically cyclical farm equipment market. This of course affected, although unevenly, both short line and full line firms. From 1971 to 1976, there were high agricultural prices and prairie farm equipment manufacturing boomed. But drought, poor yields, and lower prices for agricultural commodities could all contribute to slower sales, as they did in the 1980s. In 1983, for example, Morris Rod Weeder Co. was producing at about 40 percent capacity because of decreased demand caused by low grain prices.[48] To bridge this downturn, most companies were forced to give serious consideration to a diversification of their lines. Leon's Manufacturing, for example, expanded into nonfarm lines, and by the late 1980s, almost half of its production was nonagricultural. Similarly, Schulte Industries found some stability in producing for a nonfarm market. In the mid 1980s, it developed a rotary cutter, useful for both farm and nonfarm needs. The success of this item provided the company with a more secure economic base.[49]

Companies that ignored the necessity of improving existing lines and introducing new ones faced falling sales. While some firms had little interest in expansion, those that did saw increased export sales as the only means of smoothing out the cycles of the farm equipment market.[50] In 1967, the four largest farm equipment manufacturers in Manitoba—Versatile, Killbery, CCIL, and Agristeel—had about 5 percent of their sales in export markets.[51] The downturn in prairie farm equipment sales in the late 1960s persuaded many manufacturers that exports to the dry land areas of the United States could turn around poor sales. However, the Prairies were still the core market for most firms by the 1970s. In 1971, a survey of prairie short line companies showed three-quarters of their sales in the Prairies, about 18 percent in the U.S. and about 7 percent in eastern Canada.[52] Some firms, however, became increasingly successful in export markets. Those in Manitoba were especially successful in moving beyond their relatively small provincial market, finding markets not only elsewhere on the Prairies but also in the United States.[53] Versatile, for example, was exporting about 65 percent of its production in 1980.[54]

The overall trend was to seek markets in the United States, especially in the northern wheat-producing states. But a number of problems were encountered in this market. Despite free trade in farm equipment between Canada and the U.S., the U.S. imposed tariffs on some repair parts and defined some equipment as not eligible for duty-free entry as farm equipment.[55] For some firms, distribution was another problem with sales in the U.S. While American laws provided that dealers could handle all brands of equipment, there were informal ways around such legislation.[56] This meant that access to the U.S. market was not easy for some manufacturers, especially in areas with heavy competition. Kirchner Machine Ltd. of Lethbridge, for example, found in the 1970s that the greatest problem in selling its sugar beet harvesting equipment in the U.S. was finding dealers. There were about 13 American firms manufacturing beet harvesters, "and the top five pick off all the good dealers in each area, leaving the balance to sell direct to users."[57]

Efforts at developing export markets received some government assistance. Provincial governments in the 1960s and 1970s funded trade missions and mounted displays at U.S. agricultural and trade shows.[58] As well, joint federal-provincial programs were developed. One was the promotional "caravan" of farm equipment manufactured by 25 Manitoba and Saskatchewan firms which travelled to the northern U.S. in the early 1970s. The program, funded by the federal government with a small contribution from the two provinces, was estimated to have created $3 million in sales for participating firms.[59]

A broader base than the U.S. market was of course ideal, and some short line prairie manufacturers actively pursued these non-North American markets. In the late 1960s, for example, Versatile exported a few tractors to France and Algeria. For the latter contract, it developed a "package program," including field demonstrations, training, and a repair parts supply system.[60] The Barber Royal Commission had recommended that Canadian firms explore markets in central Europe and Asia, where some Canadian exporters had been successful.[61] Aside from the U.S. market, however, the most consistent success was found in Australia. In 1978, Versatile set up a subsidiary there, which also distributed Morris tillage equipment.[62] The Australian market was especially attractive since prairie dry land equipment required little or no redesign to meet local needs. Yet, whatever the attraction of foreign markets, domestic sales remained central to all prairie companies. As was noted in 1979 about Versatile, exports were highly important, but the western Canadian market was its "bread and butter."[63]

A company's success in the domestic market was essential for its long-term stability, and maintenance of this market share depended in part on continual research and development. Yet research and development was always a problem for small manufacturers. While they depended on design innovation for their existence, in order to attract customers they often had to price equipment at a relatively low level. This often left them with little revenue for research and development or the engineering talent similar to that found in the largest firms. Thus, while the overall product concepts used by short line firms were often excellent, refining them was more difficult. As was noted in 1964, this dichotomy was demonstrated in the tests carried out by the AMA. The recommended modifications for equipment of large manufacturers were usually found to be minor, while those for small manufacturers' equipment "have been at times quite extensive."[64] Some short line firms recognized their need for assistance in areas like design, stress analysis of machinery parts, and advice on properties of metals.[65] In one such case, Morris Rod Weeder Co. developed a prototype for a chisel plow, which the University of Saskatchewan put through a computer simulation to test the strength of stress points. While this identified weaknesses, it also showed areas where material could be reduced to save manufacturing costs.[66]

University programs like this were only one source of assistance and in some respects became less important as other institutions provided similar services. While research of a more general nature was increasingly carried out by a growing number of agricultural engineers at prairie universities and research stations, further expansion of these programs was needed to keep pace with research needs.[67] Other agencies often took up this slack. The Manitoba Research Council was established in 1963 to promote scientific research in areas of concern to Manitoba's development. DREE also provided funds for research and development, and public agencies such as the AMA and PAMI helped manufacturers on a fee-for-service basis in the design and development of equipment and in solving specific technical problems. The testing carried out by these agencies also made an indirect contribution to research since the test recommendations provided a design service for existing equipment. What remained puzzling to some observers was the reaction of some manufacturers. Some failed even to inform themselves about these opportunities, others neglected to take advantage of them, and yet others viewed them with a measure of contempt as being impractical.[68]

Although research and development was an ongoing area of concern to prairie short line manufacturers, so too was a continuing handicap in the distribution and marketing of their products. Good dealers not tied up by the full line companies were difficult to secure since full line domination of the distribution system remained significant after 1945. By the 1960s, however, dealers had gained some leeway and were able to handle more short line equipment than they had in the past. This was especially so with small equipment lines that did not compete with full line products. Dealers were no longer able to survive by selling only one line of equipment; the territory they served had increased greatly, and the market was often weak. Thus, as Roy Robinson noted in 1968, the position the full line companies now took was "much less tough than it used to be."[69]

There was, however, a limit to the number of lines that any dealer could reasonably handle, an especially important factor given the decline in the number of dealers in the 1960s.[70] Problems also remained in finding dealers for short line equipment that competed directly with main line products. Versatile encountered such difficulties when it began manufacturing swathers, combines, and tractors. This equipment was central to the full line companies, and they ordered their dealers not to handle Versatile lines or take Versatile equipment in trade.[71] In addition, the full line firms had maintained a number of other advantages. They continued to offer credit as well as lower interest rates and cash rebates during periods of slow sales. These programs required capital and an administrative system that short liners rarely had. Degelman Industries, for example, provided no financing to purchasers of its equipment. Like other short liners, it found such approaches too difficult to administer and finance.[72]

Since distribution was an ongoing concern, techniques changed in response to shifts in the market and different approaches were sometimes used for different markets. In 1971, over one-third of prairie short line firms sold part of their production directly to customers, but this was often a difficult and expensive process and most firms avoided this method if at all possible. Many firms used salesmen. Smith-Roles, for example, used farmers who had used its equipment and could speak knowledgeably about it to sell on commission. It also used a mail-order system for sales. But all firms placed equipment with dealers if possible. Some firms also used what they called "district managers" to call on dealers to promote their products and assist with sales and servicing. This was in some respects similar to the blockmen used by the full line industry. In other cases, especially in markets outside the Prairies, manufacturers used private distributors who worked on a consignment or commission basis. It was not always easy, however, to find distributors who provided good service and exerted some effort on behalf of a small manufacturer. As Ezee-On Manufacturing of Vegreville noted in 1972, good service was hard to find, although it was "very easy to get a distributor who will insert your sheets into his already six inch thick catalogue."[73] Others took a more direct hand in distributing, especially in foreign markets. Both Morris Rod Weeder Co. and Degelman Industries maintained branch warehouses in the United States, in effect acting as their own wholesalers in that market.[74]

Short line firms also used a variety of advertising techniques to maintain and expand their market. Advertisements in the farm press were consistently used. As well, in the 1970s, Morris Rod Weeder Co. was still using the field demonstrations that had proven so successful in the early years of the company. In 1972 alone, it sponsored demonstrations at 41 Manitoba locations.[75] Fairs and exhibitions continued to provide occasions for making the public aware of short line equipment, and new opportunities emerged. In 1978, the first Western Canada Farm Progress Show was held in Regina, and it became an important showcase for small manufacturers.[76]

Of the prairie farm implement firms, only Versatile went beyond the typical short line distribution methods. Initially, Versatile marketed through the full line dealers, and some observers credited Versatile's solid earnings in its early years to its low distribution and sales costs. By the 1960s, however, Versatile's greater production and sales brought with them the need to emulate the full line system. As the company's lines increased, it was forced to develop its own dealer network. Beginning in 1963, it began floor planning new and second-hand equipment, demonstrating the demands created by its extensive product mix and larger size. This in turn required more personnel to service and control a growing and increasingly complex network. Yet even during this expansion, the company for a time was able to maintain relatively low distribution expenditures because it concentrated on a relatively restricted geographical market in Canada and the northern United States. By 1970, Versatile was said to have the highest number of franchised dealers of any farm equipment company in Canada, and it was moving towards full line distribution tactics.[77] By the early 1980s, it had about 600 dealers and five parts depots in Canada and the U.S. Following its sale to Ford New Holland, Versatile acquired access to Ford's extensive North American and overseas dealer network.[78]

The contrasting success of prairie short line firms before and after World War II was the result of significant changes in prairie agriculture, in the region's industrial infrastructure, in methods of fabrication, and in the availability of component parts. In each phase, manufacturers had experienced many similar challenges. Some of these were resolved after the 1940s and were reflected in increased activity, while others continued to act as stumbling blocks. Problems with distribution endured, as did limitations on research, but a changed infrastructure for manufacturing created more favourable conditions. So too, government programs sometimes provided opportunities for overcoming other problems such as financing, training, and marketing that had seemed insurmountable for prewar manufacturers. Throughout, however, there was a basic continuity in the dynamic of the industry. G.N. Denike of the Swift Current experimental farm recognized this in 1943 when he observed that "new types of machines" and new designs were "seldom, if ever, found in the hands of larger producers of equipment." As he noted, "a study of the small, independent manufacturing plants, [and] private machine shops in agricultural areas . . . will reveal the trend for the next 10 to 15 years."[79]

This vitality had characterized the craft tradition of farm equipment fabrication before World War II. And while the isolation of this craft tradition was lost during the growth and change that took place after the war, its vitality and inventiveness were retained, even though firms were adopting a more industrial system of organization. These changes were best reflected in the growing recognition of the extent of interdependency within the industry, a development typified by the creation in 1970 of the Prairie Implement Manufacturers Association to represent the industry.

Organizing the Industry: The Prairie Implement Manufacturers Association (PIMA)

During the dramatic growth in implement manufacturing on the Prairies in the 1960s, the number of manufacturers producing farm equipment exclusively rose from 27 to 63 within a decade. The farm equipment industry was unorganized at this time, although there were a number of industry organizations to which manufacturers belonged. In Saskatchewan, a welders and blacksmiths association had served some short line manufacturers since the 1920s, but it began to lose momentum in the 1950s, probably due to increased rural electrification, which changed welding dramatically.[1] Some manufacturers also joined the various prairie wholesale and dealer associations, such as the Manitoba Wholesale Implements Association, which had been active since the 1880s. Although it was primarily concerned with the wholesale trade, manufacturers were becoming significant in the association by the 1960s, perhaps because there was no regional group to represent their particular interests.[2]

Other organizations of a national or international nature also served some prairie manufacturers. Some short line firms belonged to the Canadian Manufacturers Association based in Toronto, even though it was primarily oriented to larger manufacturers.[3] Some also belonged to the American-based Farm Equipment Manufacturers Association (FEMA), an international organization designed to promote the interests of short line farm equipment manufacturers.[4] Yet others belonged to the Ontario-based Canadian Farm and Industrial Equipment Institute (CFIEI), which was formed in the mid 1960s. It aimed to give representation at a national level to a broad range of manufacturers, although many of its members were full line companies.[5]

While these organizations continued to attract prairie farm implement manufacturers as members, the need for an association to represent and promote the special needs of this group was widely recognized by the late 1960s. The province of Manitoba noted in its submission to the Barber Royal Commission on Farm

Machinery that while the short line industry was highly competitive, co-operation in areas of export promotion, government liaison, management training, research and development, and recruitment of labour would be valuable in increasing the industry's operations and profits. It recommended that manufacturers in that province form an industry association to carry out "non-competing activities that will further the interests of the whole industry and enhance the contribution it makes to Manitoba's prosperity."[6]

By the 1980s, short line manufacturers had less difficulty in finding dealers who also handled full line equipment. This Massey-Ferguson dealer in Morinville, Alberta, also sold Morris Rod Weeder Co. Ltd. equipment in 1984. (Archives Collection, A12873, PAA)

These needs were increasingly recognized by others involved with the industry. Bart Drope, an industrial development officer with the Saskatchewan Department of Industry and Commerce, foresaw the potential advantages of such an industry organization, and he brought together a group of concerned individuals to discuss its formation. The first of several meetings was held in late 1969, and 15 manufacturers, together with representatives from IPSCO and the federal and Saskatchewan departments of industry, trade and commerce, discussed various alternatives. Initially, an association made up only of Saskatchewan manufacturers was proposed, and a steering committee of Floyd Rousell, Fibro Industries, Regina; Ross Giles, Degelman Industries, Regina; Roy Swanson, Swanson Industries, Saskatoon; Alan Scharf, Saskatchewan Research Council, Saskatoon; and Bart Drope was formed to pilot the creation of the association. Recognizing that farm implement manufacturers across the Prairies had similar concerns, the organization was very soon broadened to represent manufacturers in all three Prairie Provinces. It was called the Prairie Implement Manufacturers Association (PIMA) and was operating by early 1970. Its first officers included some of those who had supported the move to organize. George Morris of Morris Rod Weeder Co., Yorkton, was elected as the first president of the association, while Wilfred Degelman of Degelman Industries was selected as the president elect. Floyd Rousell became Secretary Treasurer; Marshall Brown, Brown Industries Ltd., Drinkwater, Saskatchewan, became program director; and Francis Doepker, Doepker Industries Ltd., Annaheim, Saskatchewan, became membership director.[7]

By the time PIMA was organized, similar associations had been established within particular industrial sectors for over a century in Canada. These associations developed from a realization that, through an industry-wide focus, members would be better able to protect, promote, and enhance their opportunities. PIMA expressed a similar ambition among prairie implement manufacturers. It created a forum in which they could communicate with each other and debate common problems and solutions.[8] To realize this objective, a broad membership base was required, and achieving a representative membership became one of PIMA's immediate goals.

While it continued to reflect its Saskatchewan origins, PIMA soon became representative of the industry on the Prairies. The first annual convention held in 1971 saw the election of a board of directors with membership from all three provinces, and maintaining this balanced representation remained a priority of the organization. Legislative, membership, and industrial development committees were organized in each province by 1973, and the first non-Saskatchewan president, Wilson Matthews, Marketing Director for Co-op Implements of Winnipeg, was elected in 1974. To ensure a continued regional balance in the organization, amendments to PIMA's by-laws in 1982 provided that the office of president had to be filled by someone from each of the three provinces alternatively. Further, the position of president elect was replaced by two vice-presidents, who would represent each of the remaining provinces at a senior level in the organization. These approaches were enhanced by a policy that annual conventions be held alternately in each of the three provinces.[9]

In addition to broad regional representation, PIMA also sought a membership that represented all facets of the industry. Initially, the main focus had been on those actually manufacturing farm implements. It was soon obvious, however, that this was too narrow a view of the industry. As a result, two categories of membership were created: regular membership for those actually manufacturing equipment, and associate membership for suppliers of parts, materials, and services, as well as those otherwise interested in the industry. The membership grew steadily, rising from 6 regular members and 1 associate member in 1971 to 20 regular members and 10 associate members in 1974. The contribution of associate members was recognized in 1989 when they gained representation on PIMA's board of directors.[10]

Early Growth

Much of the early growth of PIMA was due to the efforts of Ivan Thue, its first general manager. Before his appointment, the association had largely been run by volunteers, but it was soon obvious that the association needed full-time attention. Thue was hired in 1975 and his wife, Jean, was also employed by the organization on a part-time basis. Until their retirement in 1985, they worked as a team to make the organization an effective voice for farm equipment manufacturers on the Prairies.

As incoming general manager, Thue believed that an increase in membership was essential to the development of PIMA. His efforts in this regard were enhanced by the positive conditions in the industry during the mid 1970s, and given these conditions and Thue's considerable devotion to the task, membership had risen substantially by 1976 to 48 regular members and 106 associate members. These increases were sustained in the 1980s, and membership had reached 202 regular members and 295 associate members in 1990. Reflecting a broadening point of view encompassing the whole of the Canadian west, British Columbia manufacturers were invited to join the organization in 1989 and obtained representation on the board of directors in 1990.[11]

Communication and discussion among its members were necessary for PIMA to realize its potential. Within two years of its formation, it had a newsletter, *Factory Action,* initiated by its then president, Clem Roles, of Smith-Roles Ltd., Saskatoon. This publication was replaced in 1975 by a monthly newsletter, the *PIMA Pulse,* which helped keep members up-to-date on issues facing the industry as well as marketing and promotional opportunities. In 1990, a quarterly magazine, the *AgScene,* was also introduced as a means of informing members. In addition to these publications, the association also issued periodic bulletins on points of interest to its members. For example, in 1980, it compiled a list of products imported by prairie manufacturers and provided the results to the associate members in the hope that "Canadian companies will respond to the opportunity" to replace imports with domestic production.[12] Similarly, PIMA surveyed its members and prepared product demand forecasts to help the industry plan its production, an especially significant issue for the highly cyclical farm equipment industry. One of the most significant of these surveys was the quarterly business survey of members, which was first published in 1978.

PIMA also kept its members informed by sponsoring seminars on current topics. In 1979, for example, seminars on the preparation of operator manuals were organized for PIMA members in Saskatoon and Winnipeg in response to concerns about product liability, an escalating problem in the United States. As well, a seminar in Regina in 1980 on offshore marketing and two marketing assistance programs in 1986—one in Manitoba and the other in Alberta—informed participating members of potential offshore markets and marketing techniques. The latter programs were co-ordinated by PIMA and funded by the federal "Think Canadian" program. Further seminars, such as those held in 1987 at Calgary, Saskatoon, and Winnipeg, dealt with the topics of manufacturing supervision, liability, and loss control.[13]

In addition, PIMA encouraged its members to inform each other. Some members ran small businesses and had only a few workers, while others were large organizations with hundreds of employees. This variety in size, skills, and experience was used to the advantage of PIMA's smaller members. By promoting co-operation, the association provided a means for passing the expertise of the large manufacturers on to the smaller players who could rarely afford to hire advisers.[14] One of the vehicles for this transfer of expertise was the annual convention where members met and discussed problems and obtained information. Conventions

The presidents of PIMA during its first decade are shown in this 1980 photograph. From left to right, Clem Roles, Lawrence Edwards, Floyd Rousell, George Morris, Ray Malinowski, Wilson Matthews, Wilf Degelman, Lorne McLaren, Bill Kinnear, Francis Doepker, Bart Drope (Government of Saskatchewan), Ivan Thue (general manager). (PIMA Photograph)

Past Presidents and Honourary Members of PIMA at the 1990 twentieth annual convention, Calgary. From left to right: Gavin Semple, Ivan Thue (honourary), Lawrence Edwards, Jean Thue (honourary), Hubert Lux, Paul Soubry, Adeleine Morris, Bill Spiers, Lorne McLaren, Ray Malinowski, Clem Kuelker, Wilf Degelman, Ken McKenzie. (PIMA Photograph)

featured speakers and panel discussions on a range of topics such as production methods, patents, quality control, government programs, financial and labour management, and export markets. The emphasis was on practical issues, such as implement painting techniques, one of the topics discussed at PIMA's first convention.[15] The annual convention also became a vehicle for informing members of the goods and services offered by PIMA's associate members. The 1977 convention featured for the first time a trade show where associate members displayed their products. By 1987, it featured 74 exhibitors.[16] The trade show created a venue where associate members could reach a better understanding of manufacturers' needs and manufacturers could learn about new materials and methods.

The annual convention also provided an opportunity for members, many of whom ran small family-oriented businesses scattered over the Prairies, to meet and socialize and develop a sense of camaraderie. Other events, such as the PIMA hockey tournament instituted in 1982, rounded out the social side of the organization. PIMA also sought to provide other nonsocial benefits. Because many members had only a few employees, they often could not obtain access to disability, pension, and other benefit programs. PIMA therefore attempted to co-ordinate the availability of group benefit packages for its members. As one example, in 1976 it arranged for its members' participation in a group life/health/pension package.[17]

Promoting the Industry

Two other objectives set out by PIMA's initiating members were crucial in the organization's future activities. These were "to foster and promote the growth and well-being of member companies and the farm machinery industry" and "to assist member companies in the area of production, marketing, and personnel."[18] One of the most important steps taken by PIMA in pursuit of these objectives was its involvement in conceptualizing and mounting the Western Canada Farm Progress Show, jointly sponsored by the Regina Chamber of Commerce and the Regina Exhibition Association. Featuring only farm equipment, the event gave both full and short line companies a chance to show and promote their products. At the first show in 1978, all but one of the full line companies and almost 200 short line firms displayed over $58 million in farm equipment. This annual show proved to be highly successful in promoting both domestic and export sales for short line manufacturers. By 1985, it was classed as one of the top five farm shows in North America, providing prairie implement manufacturers' products with wide exposure to local farmers and international distributers and buyers.[19] In light of the problems they typically experienced with distribution, such venues were important for short line manufacturers, and PIMA maintained an ongoing involvement with the show. Ivan Thue, PIMA's general manager, served as vice chairman of the organizing committee in 1978 and as chairman in 1980 and 1981, and Ron Zimmer, who had succeeded Thue as PIMA's general manager, was the show's chairman in 1990 and 1991.[20]

Much of PIMA's efforts to promote its members was directed at the development of export markets. One resolution of its 1973 annual convention pressed for

PIMA was involved in the Western Canada Farm Progress Show, Regina, shown here in about 1980. The show displayed a wide range of farm equipment. (Regina Exhibition Park Photograph)

federal government assistance to "Canadian companies who exhibit their products at foreign fairs and trade shows." And in 1975, PIMA co-ordinated promotion of prairie manufacturers in the California market. In the early 1980s, PIMA was still actively pursuing the matter when its export review committee presented the first brief from the private sector to the newly formed federal Export Trade Development Board.[21] PIMA worked in several other ways to promote its members' products. In 1977, it introduced an annual directory of members and their products and in 1982 began distributing it to each of Canada's trade commission offices. In an attempt to make it more effective, the associate members committee initiated a goods and service index in the directory in 1990. Through a joint program with DREE in 1984, PIMA was able to provide its members with the opportunity to make videos of their equipment for promotional purposes. Fifty-three companies took advantage of the program, and reflecting how central dry land equipment is in the production of prairie manufacturing firms, a general promotional video, "The Tools of Dryland Farming," was also made. Another program was co-ordinated by PIMA in 1990, and 125 company videos were produced at a cost of about $1.4 million, just over $1 million of which was contributed by the federal Western Diversification Fund.[22]

Other PIMA programs also broadened opportunities for members. With $500,000 in funds provided from the federal Western Diversification Program, PIMA undertook the Innovations Project in 1988. This three-year pilot project sought to match inventors and their ideas with manufacturers able to develop and bring these inventions to commercial production, therefore providing a solution to a traditional problem on the invention-rich Prairies. Inventors and manufacturers were interviewed and registered on a computer data base, and by 1989, at least 12 inventors were matched with PIMA member manufacturers. By the latter part of 1990, over 40 inventors had received assistance in getting their ideas produced and marketed, and over 700 people had been advised in the areas of patents, licenses, research, and market and product development. While the development and commercialization of inventors' products were the main consideration, the program also assisted PIMA members by bringing new products for development to their attention. This addressed one basic impediment to technology transfer, and the program was strengthened by its affiliation with similar programs in British Columbia and Ontario.[23]

Another expression of PIMA's interest in broadening its members' opportunities was its promotion of dialogue with universities and other educational and research institutions in the region. At its first convention, PIMA had invited speakers from the agricultural engineering department of the University of Saskatchewan and the Saskatchewan and Alberta Research councils to address the membership.[24] The next year, PIMA appointed two representatives to an advisory board established under a program of the Division of Control Engineering at the University of Saskatchewan. This program concentrated on developing new products at the university, which could then be licensed and produced by those in the industry.[25] While PIMA drew on ever-widening sources of information over the years, it continued to draw upon the expertise of the universities and other educational institutions. It also worked to influence their curricula to reflect the industry's needs. As a significant consumer of skilled labour, especially production welders, PIMA lobbied for a greater number of pertinent courses and more on-the-job training programs in technical institutes and colleges in the three Prairie Provinces. In this connection, PIMA representatives served on the advisory board of Saskatoon's Kelsey Institute of Applied Arts and Science and had input on course content at other institutions.[26]

Another of PIMA's main objectives was lobbying government in the interests of the industry. In this, it allied itself with organizations sharing similar interests. It became affiliated with the Ontario-based CFIEI as early as 1973, gaining access to CFIEI's research, as well as representation on a national level. In 1987, both organizations formed a committee to develop effective techniques for lobbying the federal government and to investigate the establishment of an umbrella group with strong federal representation. The common interests of both groups were recognized in 1989 when PIMA named two people to the board of CFIEI. Such co-operation was broadened in 1990 when CFIEI spearheaded the formation of a "coalition" of industry interests. The coalition was made up of nine associations representing members involved in the manufacturing or distribution of farm implements in Canada and abroad.[27]

PIMA and Government Policy

PIMA's affiliation with CFIEI was largely an effort to maintain effective representation of the industry with the federal government. The impact of government action on the industry had been a long-standing concern of prairie manufacturers, and PIMA worked to bring the needs of small manufacturers to the attention of government and to create a unified voice for the industry in dealing with all levels of government. One of the most important areas of concern was the impact of provincial agricultural implement legislation on PIMA members.

In the early 1970s, industry concerns increased in Saskatchewan because of the proposal for an agricultural implement board to investigate farmers' complaints about warranties, availability of repair parts, and service. This board was to mediate disputes and render decisions on a new scheme of compensation to farmers for delay in delivery of repair parts. A compensation fund from which the board could pay awards was to be established, eventually being funded wholly by proportional levies on distributors of farm implements in the province. Other provisions set requirements for the labelling of machinery with date of manufacture and sale information.[28] PIMA lobbied energetically on these issues. While its views were reflected only in a small way in the amendments that were introduced, a PIMA representative was named to the agricultural implement board to represent manufacturers. Such boards were also established in Alberta and Manitoba, and PIMA representatives were named to these boards in 1975.[29]

While farm implement legislation affected equipment manufacturers in their dealings with farmers, it also touched on their relationship with dealers. The legislation of all three provinces contained buy-back provisions that required manufacturers in certain circumstances to buy back unsold equipment from dealers. PIMA was alarmed that these provisions could be interpreted in such a way that a dealer could terminate a contract simply because of being overstocked.[30] Many small manufacturers were especially concerned since "a big dealer could easily have put a small manufacturer out of business by returning a large amount of paid-for equipment and demanding repayment."[31] In meetings arranged to negotiate changes acceptable to both sides, PIMA was able to get the agreement of the Alberta–British Columbia dealers association but not that of the association representing dealers in Saskatchewan and Manitoba. While many dealers were happy with the provisions because they provided dealers with some "security in business," many manufacturers were still worried about potential abuses and continued to ask for changes to these provisions.[32]

Because it represented manufacturers in the three Prairie Provinces, PIMA worked to standardize and unify business conditions across the region. Since different standards in each province created difficulties in distribution, PIMA promoted uniform farm implement safety regulations in such areas as safety guards, roll-bar protection, noise-suppression control, and emission control.[33] As well, it worked to implement uniform commercial standards across the Prairies. This was of major concern to small manufacturers who had no effective means of protecting their equipment once it was in the hands of dealers. Many dealers gave a blanket security on all assets to the full line companies whose equipment they handled. If a dealer's business faltered, the full line firm moved against all the equipment on the premises of the dealer, including the unpaid-for equipment of the short line manufacturer. Given this possibility, PIMA lobbied for legislation that would provide small manufacturers with an effective means of protecting their equipment.[34] For this reason, it strongly supported the introduction of personal property security legislation in Saskatchewan in 1979.

In order to equalize the competitive advantages of all small manufacturers across the Prairies, PIMA also lobbied provincial governments to implement comparable tax structures. For example, it recommended in 1979 that Manitoba remove the sales tax on electricity and gas consumed in implement manufacturing. This would then parallel conditions in Saskatchewan, where sales tax was not levied on energy used in manufacturing, and in Alberta, where the province did not levy any sales taxes. These differing provincial taxes were an ongoing concern, and in a brief to the Manitoba government in 1983, PIMA once again pointed out the inequity of the effect of the sales tax on Manitoba manufacturers, as well as that of the province's payroll tax, which was not levied by the other Prairie Provinces. Provincial estate and gift taxes were also one of PIMA's earliest concerns. Many PIMA members ran small family businesses, and the transfer of the business from one family member to another, on death or otherwise, could have serious tax consequences. From the early 1970s, PIMA lobbied the Saskatchewan and Manitoba governments for removal of these taxes.[35]

Differing provincial regulations on workers' compensation were another example of varying standards within the region. The issue was first raised in 1977 in Saskatchewan because of the high rates assessed against farm implement manufacturers, and PIMA succeeded in obtaining a separate assessment category for such manufacturers in 1979. Rates immediately declined from $3.25 to $3.00 per $100.00 of payroll, and by 1984, the rate had dropped to $1.50. By 1985, efforts to have the same rates implemented in the other provinces were being realized—Alberta offered a separate assessment group for implement manufacturers at a proposed rate of $2.50 instead of the existing $4.35. In Saskatchewan, progress had been made in other ways as well. A merit system was introduced in 1979 and a majority of members in the industry were receiving rebates by 1981. In 1985, in order to maintain this safety record, PIMA co-ordinated the hiring of a safety officer to visit and advise members on safety in the workplace. In Manitoba, an agreement negotiated between PIMA and the Workers Compensation Board in 1990 saw the creation of a new classification for member firms, the implementation of a safety program, and the hiring of a safety officer.[36]

Transportation was another key area of concern because many prairie short line manufacturers delivered their products and repair parts to dealers and farmers by truck. Lack of uniformity among provinces on matters such as load restrictions and permit requirements therefore caused ongoing difficulties.[37] In 1976, PIMA organized a seminar at which its members could discuss these transportation problems with representatives of the three provincial governments. Since there was often a community of interest in transportation matters, PIMA also co-operated with other organizations. In 1977, PIMA, the Saskatchewan Wholesale Implement Association, and the Saskatchewan–Manitoba Implement Dealers Association submitted a joint brief to the Saskatchewan government that opposed lower load limits on secondary roads, which would impede delivery to rural areas.[38]

Another co-ordinated effort came in 1980. Short line companies saw their interests challenged when the Saskatchewan Advisory Council recommended legislative changes that would have restricted the rights of private carriers to the advantage of commercial ones. This was a concern because penalties imposed under the province's implement legislation, which could be imposed if emergency parts were not delivered to farmers within a certain time, did not directly apply to commercial carriers. PIMA had expressed concerns on this matter in 1977 in a brief to the three Prairie Provinces, and these proposed changes worsened the problem.[39] PIMA and representatives from the wholesale and dealers associations met with Saskatchewan government officials to oppose the implementation of the recommendations.

Another pressing transportation problem arose because all three provinces restricted manufacturers from hauling any equipment other than their own. This meant added costs since trucks had to return empty. PIMA acted with wholesale and dealers associations to remedy the problem, and this was achieved in 1986 when agricultural equipment (including parts, except in Manitoba) was exempted from pertinent legislation in all three provinces.[40] While PIMA directed its attention to these and other issues as they arose, it also consistently lobbied for total deregulation in the transportation field.

Deregulation and uniform application of legislation were also ongoing goals of PIMA in the area of equitable product treatment in both domestic and foreign markets. One difficulty in uniform application of legislation lay in Canada's excise classification system. Although agricultural products formed a category of goods exempt from federal sales tax, many manufacturers had difficulties in convincing excise officials that their products, especially newly developed ones, fell within the exempt category. By working with government on a case-by-case basis during the mid 1970s, PIMA obtained exempt status for some previously taxed items.[41] Generally, however, PIMA continued to press for a complete overhaul of the classification system. Following input by PIMA and other industry officials, a new classification system was finalized in 1981, creating one that had the flexibility "to permit easy classification of new products as they come onto the market."[42]

Because many prairie firms exported to the United States, much of PIMA's attention focused on regulatory problems in that market. While the Canadian market was entirely open to U.S. manufacturers, Canadian-made farm machinery and parts sold into the United States during the 1970s were duty free only if they appeared on the U.S. duty free list. Otherwise, varying duties were charged—manufacturers had to pay 5 percent duty on "dozer blades, front end loaders and hydraulic cylinders" and 10 percent on other items such as "shaft monitors, grain monitors, moisture testers and similar electronic equipment designed and built solely for use on farm implements."[43] Some relief came with the GATT negotiations of 1979 when it was agreed that tariff reductions would be phased in from 1980 to 1987 in respect to some items that were shown to be for agricultural use.[44] Even so, some duties still applied and PIMA continued to work for unrestricted access to the American market.

Since PIMA was a proponent of free trade with the United States, it lobbied in favour of the implementation of the Canada–U.S. free trade agreement. In 1986, PIMA's president, Adeleine Morris of Morris Rod Weeder Co., Yorkton, and its general manager, Ron Zimmer, prepared and circulated a brief in favour of free trade in agricultural equipment between the two countries. In 1987, PIMA showed its further support for free trade in its presentation to the Standing Committee on External Affairs and International Trade. While free trade in major agricultural equipment had existed from 1944, PIMA's presentation pointed out that Canadian farm implement manufacturers still were experiencing difficulties with the U.S. classification system and were being forced to pay duty on certain items.[45]

While PIMA was happy to see implementation of the Canada–U.S. free trade agreement, its joy was short-lived. In 1989, the United States placed a special duty on "a variety of traditionally tariff free agricultural equipment products." PIMA's lobbying on the point was not yet over, and it joined a "strong lobby . . . initiated to pressure the Canadian and U.S. Governments into removing the duty."[46] This lobby was the coalition of industry interests co-ordinated by CFIEI. Some progress had been made by Canadian officials on this issue by 1990 when the United States consented to certain changes. It agreed to accelerate elimination of tariffs on certain Canadian-made agricultural equipment and to simplify the system by allowing certain products to enter duty free on presentation of a certificate stating the equipment was for agricultural use.[47]

While these trade issues were of major concern, PIMA addressed a myriad of other issues on behalf of its members over the years. The application of uniform warranty requirements was an ongoing concern, as were a number of labour issues such as union certification procedures.[48] The introduction of the Goods and Services Tax in 1990 also saw the beginning of PIMA's lobbying for zero-rating of agricultural implements.[49] The variety of concerns that PIMA expressed over its first two decades reflected the nature of the prairie short line industry and many of the problems that it faced. By formulating the views of the industry on these issues into coherent and co-ordinated positions, PIMA gave added impetus to the emergence of an industrial sector on the Prairies.

Many of the issues PIMA grappled with—distribution, technology transfer, manufacturing skills, transportation, export markets, and government policy—were ones that had faced the short line industry for almost a century on the Prairies. The prairie industry's development had been limited before World War II because of the strength of eastern Canadian and U.S.-based full line companies in combination with regional and national economic and political conditions. While there was some manufacturing before 1945 on the Prairies, it was only after the war that substantial growth was experienced. The greatest strength of the industry lay in its response to regional needs through the use of prairie technological innovations in equipment. Prairie manufacturers became increasingly active in all areas of farm equipment: power, tillage, seeding, harvesting, and haying equipment. In all cases, the industry was forced to respond to a variety of factors, including the nature of the farm equipment industry, the region's economy and its infrastructure for manufacturing, and ongoing changes in the technology and methods used in farming the Prairies.

–PART TWO–

Sources of Power for Prairie Agriculture

The European settlement of the prairie west took place at the same time as significant changes were occurring in the economic and demographic nature of the North Atlantic world and in methods for working the land. Before the late eighteenth century, agriculture had been dominated by hand tools such as hoes, rakes, and flails. Although farmers also had used machines, agriculture was transformed by mechanization, especially after about 1850. The historian R. Bruce Shepard concluded that because Canadian plains settlement coincided with increased mechanization of agriculture, it "could be called a mechanized agricultural frontier." Its evolution contained "three incremental steps: up to the end of World War I, the twenties, and during World War II." The first step included animal, steam, and gas power before 1918; the second started with the adoption of the gasoline tractor in the 1920s; and the third step began when these, and additional sources of power, became universal on the Prairies.[1] Mechanization, or the use of machines that could be powered from a variety of sources, has transformed prairie agriculture and continues to do so.

Oxen, the earliest draught animals used on the Prairies, were a common source of power for farming in Manitoba in the 1870s. They were, however, usually replaced with horses—especially Clydesdales—which delivered more power although they were more expensive to keep.[2] They also fit better with the new machines that were coming into use. By the late 1880s, Manitoba farmers were using binders, seed drills, and threshing machines. Many straight grain farms, such as those on the Portage Plains, were mechanized by the early 1880s, and by the 1890s, even new settlers in Manitoba typically started homesteading with at least a seed drill, a gang plow, and possibly a binder. While most of these machines were horse powered, steam traction engines had appeared by the 1880s on Manitoba

Horses, such as this fine four-horse team seeding near Brandon, Manitoba, were an important source of field power before the 1920s. Not dated. (Harry Pollard Collection, P621, PAA)

grain farms, largely for use in threshing. Although these advances were important, further change occurred with even greater use of steam power and, after World War I, with the adaptation of the internal combustion engine for farm use.[3]

Despite these radical developments in sources of power, human labour remained crucial in farming. The rhetoric surrounding the changes resulting from the use of steam and internal combustion engines, and later of electricity, often obscured the continuing importance of human labour. Even though labour's relative contribution declined dramatically, especially in grain farming, the labour of the farmer and the farm family remained a significant factor in the cost of production. In 1948, it was estimated that 60 percent of all farm work in the United States was done by hand, and the same can likely be said of the prairie west, especially in livestock and mixed farming operations. In the same year, the livestock industry, for example, was still not mechanized, and 75 percent of work with livestock was done by hand.[4]

The use of mechanized sources of power on prairie farms was never an even process. Periods of increased use were followed by periods of minor change or even reversal. Due to a variety of technological and economic factors, a substantial increase in the use of tractors in Manitoba during World War I was followed by a decline and a reassertion of horse power. After World War II, the pace of overall mechanization increased. During this phase, horses were entirely replaced with tractors, combines took the place of threshing machines, and new equipment like hay balers appeared. All served to speed up farming and substituted machine for animal and human power.[5] In the Prairies, the pace of postwar mechanization was fastest in Saskatchewan. Between 1945 and 1953, the retail value of new machinery purchased rose from $22 million to $98 million. The province had become a major market in Canada for farm machinery.[6] Everywhere in the region, mechanization was occurring most quickly on larger grain farms in the prairie areas and more slowly in the park-belt areas and on smaller and mixed farms.

Even with mechanization and the use of animal and engine power, human labour remained important in farming as shown in this photo of stooking near Dundurn, Saskatchewan. Not dated. (Courtesy of the Saskatoon Public Library, Local History Room, LH 3096)

Greater speed and reduced labour were the main reasons for mechanization, which offered clear advantages in meeting the prairie environment and the economics of prairie farming. Increased production was seen as the way to make prairie farming profitable, and mechanization clearly stimulated higher production. Because of the short growing season and the increasingly important advantages of scale, prairie farming was always a delicate juggling act. Farmers everywhere and at all times have responded to the seasons, but a short growing season was especially relevant on the Prairies. The recollections of Philip Minifie, a Saskatchewan homesteader before World War I, expressed this precisely when he described the "ruthless pressure of time that bears down on the shoulders of every farmer and particularly every new farmer." Plowing could not begin until the ground thawed, then seeding had to begin as soon as possible, but if too early, the new growth could be damaged by frost, or if too late, damaged by frost in the fall before it had a chance to ripen. Harvesting too was a race with time to beat frost, rain, and snow. In between, the farmer summerfallowed, put up hay, looked after livestock, and completed farmyard construction, repair, and other tasks.[7]

Given these conditions, mechanization in general, and greater sources of power in particular, were extremely attractive. But, most farmers could not mechanize all aspects of farm life, and until the 1950s, most concentrated on acquiring greater power sources for field operations than for other farm tasks. For example, while farmers during the interwar years eagerly purchased tractors, they were not so quick to buy trucks to assist their farming operations. Instead they bought cars, which served their social needs for mobility. There were about 3,300 trucks on Saskatchewan farms in 1926, and by 1951, there were 50,000. In the same period, however, the number of tractors increased from 26,700 to 106,700.[8]

Steam Power

One of the most far reaching technological developments of the Industrial Revolution was the harnessing of steam for power. By the nineteenth century, it had become an accepted and widespread source of power for industry and transportation, but it was only beginning to be used for agriculture. After 1870, with the development of the differential gear and the friction clutch, steam traction engines began to appear. These engines, like the internal combustion engines that followed them, were one way the technology of the Industrial Revolution was applied to agriculture.

After 1875, developments in steam traction engines were rapid. These machines played a significant role in prairie farming, especially in breaking land and harvesting. Coal, wood, or even straw was burned to heat the water for production of steam. The earliest steam traction engines were moved by horses, but self-propelled versions were soon available. Steam-driven threshing machines and plows pulled by steam traction engines followed soon after. Steam-powered threshing machines were operating at Portage la Prairie as early as 1874, while others were working at Moose Jaw in 1884 and Edmonton in 1885.[9] These were huge machines, and some observers believed that this was not merely because they needed a large boiler. Their size also confirmed the belief that the machine represented progress, and as one commentator recollected, these "giants with plows to match added to the romance of breaking prairie sod. They almost filled the place of the railway locomotive in the imagination of the small boys, and some larger ones" as well.[10]

Steam traction engines were not simple to operate; both starting them and maintaining them were complicated. When used for threshing, they were connected by belt to the threshing machine or separator. Threshing done this way took a great amount of labour, but it was nevertheless faster and cheaper than available alternatives such as animal treadmills. Accordingly, by the turn of the century, "steam-powered threshing was seen as the only way of completing the prairie harvest." When they were used for plowing, their considerable labour needs were offset by the 25 to 30 HP delivered, and they proved to be as "economical as animal ploughing." In 1903, cultivation with horses was estimated to cost about $8 per acre, while plowing with an engine could be done for $6.40 per acre. A plow pulled by a steam traction engine could complete 25 to 30 acres per day, and as David Spector has observed, steam power had consequently become "a fact of farm life before 1910." Most of this work was done on a custom basis; steam traction engines were expensive and took a great deal of labour and skill to maintain and operate.

Operators had to obtain a license after completing written hand-in examinations. Owning such an engine was clearly beyond the means and abilities of the average farmer. Steam traction engines also had certain drawbacks: fuel was expensive, water for the boilers was often difficult to obtain, and, in plowing, the soil had to be perfectly dry before the heavy engines could be driven on it. Sparks from steamers also caused many fires, and while steam engines were generally reliable, the boilers sometimes exploded, killing or injuring workers. Clearly, the arrival of the internal combustion traction engine offered several technological and economic advantages over steam traction engines. An added advantage was that anyone could (or could try to) operate them since an operator's license was not required. Facing this challenge, the steam era had to all intents and purposes ended by World War I.[11]

Tractor Farming

Steam power helped to convince prairie farmers that mechanized sources of power were indispensable, and prairie agriculture before World War I developed within this framework. This view was later confirmed with the adaptation of the internal combustion engine for agricultural use. While the first internal combustion engine was developed in the late 1870s, it was not applied to field work until 1901–02 when the Iowa inventors C.H. Parr and C.W. Hart brought out an internal combustion engine tractor. Copying the size and styling of steam traction engines, these early gas tractors were cumbersome and huge, often weighing between 10 and 25 tons.[12] In part, this resulted from tradition and fashion since "the public wanted them big," but large size was necessary for some field operations. In an age when mile-long furrows were the rule (reaching two or four miles in some places), high power was needed.[13]

In about 1905–06, a farmer near Clearwater, Manitoba, bought what was probably the first gas tractor (a Kinnard Haines) for farm use on the Prairies. By 1908, there were about 400 gasoline tractors on prairie farms. In 1906 alone, about 11 companies had begun manufacturing gas tractors in North America, and at the Winnipeg Tractor Trials of 1908, the competitors included five gas tractor manufacturers as well as five steam traction engine manufacturers. A gas tractor produced by Kinnard Haines won the prize, which was "the first indication of the supremacy of the gas tractor over steam." Until this point, farmers had little basis for judging the claims of manufacturers, and the Winnipeg Tractor Trials, which began in 1908 and continued until 1912, were intended to alleviate this problem. By 1911, there were about 4,000 gas tractors and about 4,800 steam traction engines on the Prairies, but within a year the number of gas tractors had more than doubled, reaching an estimated 9,000.[14]

By this point, gas tractors were competing directly with steam traction engines. They could develop as much horsepower, and they were easier to start. As well, they combined functions; useful for all sorts of field work, they could also be used on roads for hauling heavy loads such as grain to the elevator. The gasoline or kerosene that they burned was also less awkward to handle than coal or wood. In addition, they did not need water to operate.[15] Apparently, gas tractors were adopted quickly in areas with poor water supply, such as southern Alberta and south-western Saskatchewan, while, as late as 1915, steam remained "commonly used" in Manitoba and the northern areas of the other two provinces with more plentiful water. Gas tractors were also used more frequently in custom work since they required less labour. Even so, these advantages were partially offset by the trouble of operating them in cold weather; an especially important consideration in the prairie threshing season.[16]

By the end of World War I, it was universally accepted in North America that the gas tractor was preferable to the steam traction engine. Light-weight gas tractors had now become the rule. The Winnipeg Tractor Trials had helped to advertise their benefits, and the Bull Traction Machine Co. produced the first popular small gas tractor in 1913. By 1915, the typical gas tractor on the Prairies was about 15 HP and ran on either gasoline or kerosene. While gasoline was more expensive per unit of fuel, it delivered more power per unit and was easier to use than kerosene, which had to be mixed with water. While tractors using kerosene were still on the market in 1920, those that used gasoline steadily gained ground.[17]

The small size of these tractors represented the triumph of a particular view of how power could be utilized. Henry Ford was a champion of light tractors, and his genius proved to have as great an impact on farming as it did on private transportation. One anecdote told of Ford finding a "kindred spirit" at the Winnipeg Tractor Trials. This supporter illustrated "his conception of the light tractor by the story of a cat chased up a tree by a dog: 'The cat,' he says, 'didn't have weight but she had traction.'" Yet the belief that weight was necessary for strength was still entrenched, and another bystander argued that "if a brick had been tied to the cat's tail, corresponding to the plows behind a tractor, the dog would have put the cat out of business."[18]

The cat in the tree proved to be the way of the future. In 1915, Henry Ford announced he would produce small tractors. They were "a radical departure from the massive steamers and early gasoline engines and marked another step in the evolution of the farm tractor." In 1917, he sold 6,000 tractors to the British government, which was looking for ways to maintain agricultural production during wartime. For the same reason, the Canadian government purchased 1,000 Fordson tractors, most of which were sold to farmers on the Prairies. In 1918, it also lifted the import tariff on small tractors valued at less than $1,400, an exemption that was extended after the war ended.[19]

These wartime programs went a long way in establishing small tractor design as the prototype adopted by the industry for the next 25 years. But they also contributed significantly to the general acceptance of the general viability of tractors in prairie farming.[20] During the war, the Canadian government had urged farmers to use tractors to increase production, and provincial governments also encouraged their use by co-ordinating orders for tractors.[21] These efforts did not, however, lead to the immediate replacement of the horse in prairie farming, and the transition period was marked by an often emotional debate between advocates of "tractor farming" and "horse farming." Many of the early tractors were mechanically unsound, which supplied critics with good ammunition. W.R. Motherwell,

Saskatchewan's minister of agriculture, was critical of the high depreciation and repair costs of tractors. In 1913, a Lethbridge newspaper used some of his criticisms in an editorial which charged that "many a farmer would be better off today if he had never seen a tractor. In the majority of cases these purchases have been ruinous mistakes. In a few short years this machinery reaches the scrap heap." [22]

The debate was also spurred on by lobbyists for each industry, each of whom produced much exaggerated publicity.[23] Prairie farmers were torn between an emotional attachment to the horse and an equally powerful desire to be modern and to take advantage of the financial benefits that machines promised. In 1920, one writer, despite his assertions of impartiality, summarized the case in favour of the horse and against the tractor:

> On the one side—hard-headed experience and long records of dependability; on the other, brilliant salesmanship, national advertising, persistent propaganda. On the one side, natural pride and affection for the knowing, living animal; on the other, the lure of 'up-to-dateness,' of power amassed into a single unit. On the one side, self-repair, self-reproduction, flexibility, long life and fuel at cost; on the other, heavy investment, expert repair and operation requirements, retail prices for fuel and rapid depreciation.[24]

Some farmers were convinced by such arguments and supported the use of 8-, 12-, and even 16-horse outfits.[25] However, this "horseman's answer to the tractor" was clearly in a losing position by the late 1920s. By this time, the major problems faced by tractor advocates were being addressed. Tractor design had improved over early models such as those first manufactured by Ford, and there were fewer difficulties with repair. Since many farmers did not know how to repair tractors, manufacturers had begun to mount tractor schools through their dealers in most towns. The schools promoted tractor farming in general, as well as particular brands of equipment, and farmers were shown new equipment and given instruction in grinding valves, replacing piston rings, and other mechanical work.

In the 1920s, the use of tractors on the Prairies began to increase rapidly. Sales were high immediately after World War I, but dropped in the early 1920s because of low wheat prices. They then rebounded in the mid 1920s, and by 1926, one in every seven farms in Saskatchewan had a tractor, the highest rate of mechanization on the Prairies. By 1931, there were about 82,000 tractors on prairie farms, with about 53 percent in Saskatchewan, 30 percent in Alberta, and 17 percent in Manitoba. Along with combines, this represented an investment of about $95 million, plus an additional $25 million for "accessories" like swathers and larger seeding and plowing equipment. However, few farmers used only tractors—most continued to use horses for some work and tractors during peak work seasons in spring and fall.[26]

Most of these tractors were small, about 15 to 20 HP, but they could accomplish a significant amount of work.[27] Eight acres per day could be plowed by a 15 HP tractor pulling a two-bottom 14-inch plow. This compared with four acres a day with the same equipment pulled by five horses. Similar gains could be had in

Tractor schools, where farmers learned to repair and operate gasoline tractors, were popular before 1945. This one, probably connected with the war effort, was held in Smoky Lake, Alberta, in 1943. (Ukrainian Cultural Heritage Village Collection, UV521, PAA)

harvesting, hauling grain to market, and farmyard tasks like grinding grain. Equally significant, tractors could be used for several field tasks at once. For example, all the work in preparing a seed-bed (cultivating, packing, and harrowing) could be done as a single operation. This saved money and labour and allowed earlier seeding, a critical matter on the short-season prairies.[28] Despite such clear-cut advantages, the adoption of tractors was a complex process. It was influenced by a number of factors, including the technological development of tractors, availability of equipment suitable for use with tractors, equipment costs, and farm income.[29]

This direct relationship to income was demonstrated by the Depression, which set back the adoption of the tractor on the Prairies and gave horse advocates a second wind. Tractor sales collapsed in the early 1930s, and the place of the horse in agriculture was reaffirmed for a time as farmers returned to the use of horses to escape the costs of running a tractor. However, tractors were making a comeback by the late 1930s. Their earlier attraction was further enhanced by improvements in design. Although it was not commonly used until after World War II, direct power-take-off had been perfected by the mid 1920s. It meant that implements could be run directly off the tractor engine rather than by ground drive. With ground drive, the moving parts of an implement (such as cutting bars on binders and mowers) were connected to the implement's wheels, which, as they turned, also powered these moving parts. Speed and consistency of operation of an implement would therefore vary with the speed of the tractor and with field conditions.

Power-take-off was an important and significant change over this method. It consisted of "an arrangement of shafts and gears connected to the transmission and terminating at the rear of the tractor so that it can be connected conveniently to any machine." Usually, the power-take-off was built into the tractor's transmission "and is completely enclosed with the exception of the projecting end of the shaft to which the connection [to the implement] is made." The power-take-off was disengaged by operation of a hand lever on the tractor.[30] This development was important because equipment could now be operated more efficiently.

This innovation was among the most significant labour-saving developments in agricultural machinery. Also popular were rubber tires, which were coming into use by the end of the 1930s. They did not pack the soil as did steel wheels, and they offered increased comfort, speed, and fuel efficiency. All of these advantages were proven in tests at the University of Saskatchewan and at the Swift Current experimental farm. In 1938, 41 percent of new tractors sold in Canada had rubber pneumatic tires, and steel-wheeled tractors were also being refitted with them. Tractor schools became common once again, and one commentator observed that many farmers, who were being nagged by their sons, bought a tractor "to keep the boys at home."[31] Tractors became even more attractive with the greater use of diesel engines. While diesel engines had been developed in the late 1890s, they were used primarily for stationary engines and ships. The technology was later adapted to farm tractors, and by the 1930s, tractors with diesel engines were on the market. Diesel was less expensive than gasoline, and given its higher possible compression, it was also more efficient.[32]

The other significant development in tractor technology was the installation of hydraulic systems, increasingly in conjunction with the power-take-off. In the 1940s, hydraulic lifts operated by levers from the tractor began replacing mechanical lifts on the equipment itself. As well, depth control on tillage equipment, which had relied on the use of levers until the 1950s, was increasingly controlled with hydraulic lifts.[33] These hydraulic systems were advantageous for a number of reasons. As one commentator noted, although they "may not be as efficient as chains or gears, they are so much more flexible in application that this factor alone offsets any drop in efficiency." Moreover, they decreased the amount of physical work and increased safety.[34]

The manufacture of new tractors, like other farm equipment, slowed during World War II. The war created extraordinary demands on prairie agriculture because of government insistence on higher production with less use of gasoline and other war essentials. Mindful of the lessons of World War I, however, the federal government warned farmers not to sacrifice soil fertility in pursuit of higher production.[35] Yet, even with these limitations, the war was a catalyst for significant change on prairie farms. Demand for increased production helped popularize the advantages of tractor farming and mechanization in general, and it stimulated a major movement toward the complete replacement of horses with motor power. One study of central Saskatchewan revealed a significant decrease in the use of horses between 1939 and 1944. This was especially so on large farms (850 acres and over), 90 percent of which used only tractors by 1944. Similarly, use of trucks, combines, and larger tillage equipment also increased.[36]

In common with other sectors of the Canadian economy, farmers had high hopes for the postwar years. Towards the end of the war, there was a high level of pent-up demand coupled with a desire for newness. Saskatchewan farmers indicated that the greatest change they planned to make was further mechanization. Topping their wish list was rubber-tire tractors. There was a general emphasis on mechanization, and almost all farmers wanted to purchase larger power units to make a complete transition from horse to motor—especially diesel—power.[37] Most of these hopes were realized after the war, which was clearly shown in the massive increase in tractor sales. In 1947, they increased on the Prairies by between 50 and 60 percent over the year before, and in 1949, almost 62,000 tractors were sold throughout Canada. Diesel engines were popular, and by the early 1950s, there had been "an almost complete changeover to tractors" in Saskatchewan. This development was not unique to Saskatchewan, and as a result, Cockshutt Farm Equipment had "practically stopped production of horse-drawn machines" in 1952 because western demand for them was so low. This rapid pace of mechanization kept up until the 1953–54 farm recession led to declining farm implement sales in the region.[38]

In 1952, the Prairies continued to lead the nation in the use of tractors. Prairie farmers, for example, had twice the number of tractors as did Ontario farmers. While there were slightly more tractors than farms in Saskatchewan, there was just under one tractor per farm in Alberta and Manitoba. By this time, most farmers would not accept anything but the tractor for field power.[39] The continual refrain in reports on farm machinery in these years was that tractors were becoming bigger. In 1965, most tractors on the Prairies were about 70 HP and a growing number were 90 HP. This was at least a doubling of size over those in use during and before World War II. These were powerful machines by any standard, but they were especially so in contrast to those in use in eastern Canada where most tractors ranged between 35 and 60 HP. While some experts foretold that 150 HP models would soon be common on the Prairies, these forecasts were modest. By 1966, four wheel drive models were on the market, including one manufactured in Winnipeg by Versatile. The demand for increased power was so great that some farmers made their own four wheel drive versions by hooking two tractors in tandem.[40]

Tractor size increased even more in the next decade. The 1970s were characterized by major growth in horsepower capacity on prairie farms, and by the end of the decade, 200 HP tractors were common on large farms. They were justified by farmers' efforts to combine field operations as much as possible to save costs and labour. The customary assumption that increases in farmers' real incomes had to come from reducing farm labour costs joined in the 1960s with a growing belief that farm size had to increase even further. By the late 1960s, the average farm in Alberta and Saskatchewan had reached 800 acres, while the farm workforce had dropped by 8.5 percent in 1966 alone.[41] By the early 1970s, a hitherto unheard of specialization in tractors was being encouraged. Its advocates contended that farmers needed a main field tractor (four wheel drive), a smaller utility tractor for a back-up for field work, and an even smaller chore tractor of around 60 HP for yard work. This was not typical, however, and most farmers had a utility tractor for field work and possibly a chore tractor for yard work. Still, the long-standing view of the tractor as a machine for all purposes was being replaced with one that espoused a capital intensive specialization.[42]

These changes in tractor power created a chain reaction through the system. Broadly, tractors promoted self-sufficiency among farmers who could now do alone what had previously required group labour. In turn, self-sufficiency in combination with increased farm size stimulated yet further technological expansion. In 1965, "the trend to bigness" in tractors was "pushing implement sales up as well. Farmers need larger implements to get the most economic use from their large tractors." While much of this growth was necessary in the context of changes in farming, some of it was not. Although many farmers would deny the assertion, prestige was also important in the emphasis on bigness.[43]

The 1980s saw an increasing emphasis on four wheel drive tractors. In addition to greater stability, better traction, and more flexible power and speed, they allowed farmers to work in wet fields if necessary. While smaller two wheel drive tractors still made up about 80 percent of the tractor market on the Prairies in 1980, larger four wheel drive machines were the way of the future. J.I. Case was making 200 and 300 HP four wheel drive tractors, now the typical size of new tractors, and in 1980, the largest unit manufactured by Versatile was 348 HP. Farm size was increasing and greater power was needed for new tillage equipment that now was as much as 60-feet wide, up from 40 feet only a few years before. Larger equipment represented fewer trips to the field and therefore less cost for fuel, even though fuel efficiency had increased by about 35 percent between 1965 and 1980.[44] The move to four wheel drive tractors was somewhat tempered in the early 1980s when the farm implement market collapsed. In 1981, sales of four wheel drive tractors in the Prairies had reached 15,000, but by 1982, sales were declining because of high interest rates. Although they picked up briefly in 1984, sales reached only 3,200 in 1986, reflecting the ongoing problems caused by low commodity prices and falling land values which reduced farmers' borrowing capacity.[45]

Farm Power and Manufacturing on the Prairies

The full line companies in the United States emerged quickly as the undisputed leaders in the manufacturing of steam traction engines and gas tractors. There were few successful attempts to manufacture these on the Prairies. Prairie manufacturers did not control the pace of technological change in tractors; a crucial element that placed them at a disadvantage. Yet manufacturing was never an isolated process—ideas developed in one part of the continent were easily applied in another part, and use of component parts (such as engines) manufactured elsewhere were combined with local fabrication. Manufacturing of any sort, including farm equipment, was rarely self-contained and exclusive.

In the steam era, most steam traction engines were produced outside the Prairies. There were, however, a couple of known exceptions. The Peterson Brothers Iron Works, Winnipeg, a company established in 1898, repaired and assembled steam traction engines. About the same time, they also began manufacturing their own steam engines. Although this would have involved a considerable degree of assembly of component parts, the company carried out enough unique fabrication to give their name to the machine.[46] The Brandon Machine Works Company Ltd. (later the Brandon Implement and Manufacturing Co.) also manufactured a steam engine. The company asserted that it was "their own make," although they probably used some component parts manufactured elsewhere. It was called the Brandon Cornell, and the company built 15 machines from 1900 to 1902 and planned to construct 30 more in 1903. Production seems to have dropped off by 1904, however, since the company was then concentrating on manufacturing small portable and stationary gasoline engines. It seems not to have returned to manufacturing steam traction engines.[47]

The "Brandon Cornell" was among the few steam traction engines manufactured on the Prairies. (*Farmer's Advocate*, December 20, 1902)

The manufacture of internal combustion engines proved to be a more promising area of activity, as illustrated by another Brandon firm, Manitoba Engines Ltd. (formerly the Manitoba Windmill and Pump Co. Ltd.). At its formation in 1903, the company specialized in windmills and pumps, but by 1905, it was producing both stationary and portable gasoline engines which were used for cutting wood, running grain grinders, and other similar chores. These engines were apparently popular; from 1905 until at least 1920, the company produced a variety of sizes, usually from 1.5 to 30 HP. In addition to the factory in Brandon, the company maintained a warehouse in Calgary. In 1915, it advertised itself as "the only manufacturing

Stationary gasoline engines were manufactured in Brandon in 1915 by Manitoba Engines Ltd. (Manitoba Engines Limited brochure, 1915, Ag. 12. II. B. 48, SABS)

concern of its kind in Western Canada." It also produced metal goods like pulleys, iron rods, and grain grinders, and it continued to manufacture water pumps and windmills, then a popular source of power for pumping water on farms. In 1908, the company asserted that it was the only windmill manufacturer west of the Great Lakes, although it is likely that their windmills ("the most scientifically designed mills the world had yet produced") were of American design, perhaps manufactured under license.[48]

The growth in the use of the gasoline tractor also created some manufacturing activity on the Prairies. Attracting a manufacturer of heavy farm equipment was a popular goal of prairie town and city boosters, who argued that the region's agricultural activity as well as its natural resources guaranteed success for such manufacturing. In part this was mere boosterism, but prairie residents also tended to overestimate the importance of their role as consumers of heavy equipment like tractors. While there were periods of high consumption, demand was not consistently high enough to sustain such manufacturing, especially in a sector that was characterized by such a high level of technological change. Indeed, as John Phillips, formerly with the Cockshutt Company, noted in 1968, tractors were probably the most difficult pieces of equipment to manufacture. An annual sales volume of between 4,000 and 5,000 units was "insufficient to support the large amounts of money required for continuous engineering improvements" and the necessary retooling of factories to keep pace with these changes.[49]

This conclusion, so true in the 1960s, also applied to the entire history of tractor manufacturing in North America where innovation in tractors was continuous. And clearly, the prairie market was relatively small in the North American context. For instance, sales of tractors in Ontario and the four western provinces in 1969 totalled

15,040 units, compared to 23,870 in the three American states of Minnesota, Iowa, and Wisconsin where most tractors were manufactured. National sales demonstrated this difference in market size even more emphatically. In 1971, 131,623 tractors were sold in the United States in comparison to 18,685 in Canada.[50] The only prairie company that emerged to challenge the dominance of the major manufacturers in tractors was Versatile, which began production of a four wheel drive tractor in Winnipeg in 1966.

Although it was the most successful, Versatile was not the first company to produce tractors on the Prairies. Immediately before and after World War I, there were several prairie companies that assembled tractors and at least one that manufactured them. These efforts were part of the ferment created by the widespread adoption of the new technology. When light gasoline tractors were first developed, about 200 companies manufacturing small tractors appeared "overnight" in North America. Although most of these companies disappeared as quickly as they had appeared, they indicated popular anticipation about the success of the gasoline tractor as a source of power and the way in which new technology stimulated manufacturing activity.[51]

In keeping with the growing enthusiasm about gasoline tractors, in 1911 the Hero Manufacturing Co. Ltd., Winnipeg, bought the patents, drawings, and factory equipment of Albert Espe, a Crookston, Minnesota, inventor and manufacturer of a 20 HP tractor. Hero planned to move the factory, as well as Mr. Espe as factory overseer, to Winnipeg. Four tractors per week were planned; they were to be called the Hero, the same as the company's other products. Financing was an obvious problem, and the company ran advertisements urging farmers to "Become a Manufacturer" by buying shares in the company. This was apparently unsuccessful, and the plan seems not to have come to fruition.[52] In 1915, yet another group attempted to establish a tractor factory in Manitoba. It appears that a prototype of a tractor called The Manitoba Universal was produced, and while it was claimed to be better than eight good horses for plowing, evidence of its production has not been found.[53]

Greater, although hardly impressive, success in manufacturing tractors occurred in Medicine Hat, Alberta, after World War I. One attempt was made by Walter Martin and A.P. Phillips who were involved in gas drilling and fabrication of drilling tools and other metal working and repair. They began assembling Stinson tractors in 1919, using the design of Charles Stinson, an American manufacturer. While Martin and Phillips advertised it as "made for the West at Medicine Hat and sold at American prices," it was not a prairie-designed machine. The Stinson tractor ran on kerosene and could plow almost 13 acres in 10 hours, considerably more than a team of horses. The tractor, despite extensive advertising, was never successful and production stopped in 1922.[54]

Another tractor produced at Medicine Hat was The Canadian 14–28 Tractor. It was manufactured by the Alberta Foundry and Machine Company Ltd. and was probably the earliest complete manufacture (as opposed to assembly) of a tractor on the Prairies. Its production was announced in 1919, and while it was designed

One of the early gasoline tractors manufactured on the Prairies was the Stinson in Medicine Hat, Alberta. Designed by an American, the tractors are being displayed on Medicine Hat's main street, 1919–20. (2003–103, GAI)

by a Minneapolis inventor, the fabrication was done in Medicine Hat. Its promoters expected to receive tariff protection, but at a cost of $1,200, the tractor had to compete with the tariff-free category of tractors of $1,400 or less announced in 1919. Perhaps this was the reason why the tractor did not appear on the market until 1920; the manufacturers were possibly waiting for the tariff to be reimposed.[55] The tractor had a 14–28 HP rating delivered by a two cylinder engine. The engine was started with gasoline and ran on kerosene. It had a wood frame and wooden spokes on the rear wheels. Advertisements claimed that it was "not a high-speed tractor encumbered with highly scientific devices. Western farmers don't need that kind of a machine." The company also promised to sell repair parts at cost plus a "minute profit." Like the Stinson, it was a failure in the market, and the company produced only 20 units before suspending fabrication. The Alberta Foundry and Machine Company nonetheless continued to be active, producing other equipment.[56]

These early attempts at tractor manufacturing were stymied by lack of capital, a relatively small local market, rapid technological change directed by larger firms, and foreign competition. Following the disappointments of the two Manitoba attempts, as well as the two efforts in Alberta, tractor manufacturing did not occur again on the Prairies until Versatile began production of tractors at its Winnipeg plant in 1966. Because of this, prairie farmers bought tractors produced elsewhere, and the tractor market by the 1960s was dominated by Massey-Ferguson, IHC, John Deere, and the White Group. Massey was strongest in the east, while Deere was especially strong in western Canada. All produced tractors of up to 90 HP, but in

1966, Case brought out a four wheel drive 100 HP tractor at a cost of between $16,000 and $18,000.[57]

Versatile also produced a four wheel drive tractor in 1966, the first to be manufactured in Canada. Its design was in keeping with the innovation and lower retail cost that Versatile was known for. It was a four wheel drive machine with 100 HP at the drawbar and sold for around $10,000, considerably less than competitive models. Declines in the farm labour force and increased acreage demanded more powerful and efficient tractors, a trend that Versatile encouraged by making this machinery more affordable. The steel came from the Manitoba Rolling Mills at Selkirk. Versatile manufactured its own transmissions but the engines were manufactured by Cummins Diesel, a major North American engine manufacturer. In 1967 alone, Versatile sold 1,200 tractors.[58]

Versatile (Winnipeg) manufactured the first four-wheel-drive tractor in Canada in 1966. (VFEO–Ford New Holland Canada Ltd. Photograph)

Although tractors were a small part of Versatile's total production, the company aggressively entered the market in the 1970s and 1980s and continued its innovative design and research. In 1970, its four-wheel-drive tractor was still considerably less expensive than any U.S. imports. The number of models Versatile offered steadily grew. In 1971, it introduced a four-wheel-drive tractor with hydrostatic drive that allowed a wide range of forward speeds. By 1973, it was producing five different models of four-wheel-drive tractors, ranging in engine power from 170 to 470 HP, and by 1983, it had eight models of tractors.[59] Its 200 series were small tractors in the range of 100 HP with conventional drive. They contained almost 20 years of design experience and thought. With articulated steering, they offered important advantages over large rigid-frame tractors which, because of their wider turning radius, were not wholly satisfactory for farmyard use or when working edges and corners of fields. Therefore, while articulated steering was not a unique development, it was an important feature of the tractors. A more radical feature was the use of a bidirectional design, which permitted the tractors to work in either direction. Because there was a power-take-off and a three-point hitch at either end of the tractor, farmers were given greater flexibility in connecting equipment to the tractor and could use either pull or push implements with greater ease.

As it had done since 1966, Versatile continued to argue that its tractors were easier to repair than competing models because of fewer moving parts and the use of modular design. This was consistent with Peter Pakosh's design philosophy that simplicity was the key to successful equipment. Versatile and the federal government each put $6 million into the research and development of the bidirectional tractors. Although Versatile's tractor production was low in comparison to that of the full line firms, it showed that economical production was possible with smaller but specialized production.[60]

Another tractor producer to appear on the Prairies during the 1970s was Terra Power Tractor Co., a subsidiary of a Romanian tractor firm. There were a number of demonstrations of these tractors across the Prairies in the early 1970s, and dealers

Versatile's small bidirectional tractor, advertised as a "push-pull" tractor, allowed the operator to attach loaders to either end of the tractor and use them simultaneously. (VFEO–Ford New Holland Canada Ltd. Photograph)

were recruited. After several years of fruitless negotiations for financial support from the provincial and federal governments, assembly of tractors began in 1973. The plant was located in Saskatoon, and by 1975, the company was assembling four models, using some Canadian-manufactured components, especially tires. The tractors were based on earlier Massey-Ferguson designs. The company specialized in the small, or chore, tractor market and was selling about 250 tractors a year in the mid 1970s. By the mid 1980s, the company announced plans to pursue sales in Ontario, the best market in Canada for small tractors.[61] All other tractors sold in Canada during the 1970s and 1980s were imported. By this time, Canadian full line companies were importing tractors made by American firms in the United States or elsewhere. In addition, CCIL imported Deutz and Volvo machines, although it did carry out some assembly in Winnipeg.

While tractors did not play a large role in farm implement manufacturing on the Prairies, they did create some spin-off manufacturing. Attachments for tractors were an important sector of prairie manufacturing. In 1918, for example, Martin Parker of Tyndall, Manitoba, developed a special hitch to avoid the common problem of "breaking the Plow or Tractor when a root or stone is hit, causing considerable repair expense as well as loss of time." At $10, Parker claimed that "you cannot afford to be without it a single day." [62] Another example of this early spin-off from tractor use was a steering device marketed by Western Steel and Iron Co., Winnipeg. The device fit at the rear of the tractor and followed the furrow, helping the operator keep a straight line without the assistance of someone to watch the furrow. Advertised as "SAVES ONE MAN'S TIME," it was an early illustration of prairie farmers' ever-present concern about labour.[63] Many similar items were manufactured over the years, and the persistence of these concerns is illustrated by the introduction in 1980 of a new hitch by Victory Equipment Ltd., Lethbridge. Designed to pull several implements simultaneously, it reflected the increase in tractor power that had taken place. However, recognizing that many farmers still had smaller cultivators that were serviceable, the hitch combined "the versatility of small machines with a wide cut." [64]

The manufacture of tractor attachments was often a risky proposition; if the item became highly popular, the tractor manufacturers began supplying it with their tractors, either as an option or as standard equipment. This was the case with tractor cabs. Cabs had been manufactured by MacDonald Bros. Aircraft Ltd., Winnipeg, as early as 1947 as part of the company's shift to peace-time production. It was not until the 1960s, however, that production of cabs grew and flourished for a brief time. They were being produced by a company in Minnedosa in 1962, and in 1965, Fibro Industries, Regina, began manufacturing fibreglass cabs. Called the Comfort Cab, the company was producing about 20 per week, but given high demand, it had doubled its production area within a year. The cabs cost $640, and radios, windshield wipers, and heaters were available as options.[65] While the future for this industry looked bright, it was soon disrupted when tractor manufacturers began offering cabs with their tractors. By 1972, when Versatile began including cabs as standard features, Fibro Industries was manufacturing 1,100 cabs a year, although it planned to begin to phase out this production because of such factory installations.[66] While a few firms continued to manufacture cabs after this point, it was a much smaller industry. The inclusion of these items as standard features demonstrated a dramatic shift in farmers' attitudes to what some critics had called "luxuries." Tractors with soft seats, radios, and cabs were described by Versatile in 1979 as necessary because "the farmer is a businessman, and these comforts are ones which most businessmen have access to."[67]

An important segment of the tractor attachment industry was loading equipment. Front-end loaders, stackers, buckets, blades, forks, and other implements were increasingly manufactured after World War II as tractors became larger and more opportunities for mechanizing farm tasks became apparent through the use of hydraulics. By the early 1950s, the value of tractor-mounted hydraulic loaders was widely recognized. As was noted in 1953, "hydraulics are taking the muscle-work out of farm work, and if there is anything that a farmer appreciates, it is convenience and economy in getting his chores done."[68] There was great variety in hydraulic loaders, which were devised for every farm task—from handling manure and feed to removing snow. Most of these used the tractor system for their power source. The earliest types were bolted on the tractor, and although connecting and dismounting the loader were said to be simple, this was rarely so.[69]

A significant development in this respect was the invention in 1966 of an easy-to-attach front-end loader by Demkiw and Yakimetz Equipment Ltd. of Vegreville, Alberta. Their system could be attached and removed from the tractor in minutes. This was in sharp contrast to former systems that had been bolted onto the tractor and took hours to mount and remove. It was a development that revolutionized the tractor loader industry and directly met farmers' concerns for convenience and speed of operations. Reflecting its new product, the company changed its name to Ezee-On Manufacturing Ltd.[70] The principles of such easy-to-attach front-end systems were quickly adopted by other manufacturers for other equipment. While such equipment had an agricultural application, it often also served an industrial market. This gave these manufacturers a more broadly based market and meant that they did not compete solely with tractor manufacturers who might begin to offer similar equipment as options with their tractors. An example of this more broadly based market was demonstrated by an increasing demand for dozer blades. Attached to the front of a tractor, they were used for moving earth and snow and similar tasks. Leon's Manufacturing Ltd., Yorkton, which began manufacturing dozer blades in the 1950s, expanded considerably with the growth of this market. Given the popularity of dozer blades for snow removal, Leon's brought out blades that featured a rippled surface to make them self-cleaning. The blades came with easy mounting and dismounting mechanisms.[71]

A more broadly based market was also available for a variety of cartage implements. As early as 1903, the Western Implement Manufacturing Co. Ltd., Winnipeg, manufactured steel-wheeled wagons. The company claimed that its wagons, which were low and wide, would allow farmers "to haul larger loads with less horse power and load them with less manpower." They also manufactured a

Dozer blades, such as those manufactured by Leon's Manufacturing, Yorkton, Saskatchewan, demonstrated increased use of tractor attachments for accomplishing various farm and other tasks. (Leon-Ram Enterprises Inc. Photograph)

wooden wagon box that combined several functions. With adjustable sides and ends, it was suitable for hay, grain, or stock, and the company described it as a "unique article, and one of the strongest and handiest of its kind and several farmers state that it fills a long-felt want, as the East was not making just what Western farmers were wanting."[72]

The increased use of trucks of course reduced the market for wagons, but they in turn created other opportunities. Especially after World War II, there was a growing market for truck boxes and hoists. These products were manufactured everywhere in the Prairies, and the market grew quickly during the 1970s when length of hauling distances began to increase because of rail line abandonment, the decline of small towns, and the closing of elevators in many points. This was a highly varied market, and there were numerous firms involved in manufacturing boxes and hoists. An equally large number of companies also manufactured other truck equipment like cab guards, grill protectors, and bumpers. The use of larger field equipment also created new opportunities. All-purpose flatbed decks, as well as more specialized trailers for hauling equipment like swathers and grain drills, were produced by numerous metal-working firms throughout the Prairies. These manufacturers were not solely concerned with agricultural equipment, but served industrial and recreational markets as well. The opportunities for manufacturing were in this way broadened by technological change in field power, transportation, and cartage.

Mechanization in general, and greater sources of power in particular, had far reaching and dramatic effects in prairie agriculture. It can be hazarded that they contributed to increased farm profitability, but they also—by making wrong tillage practices possible—contributed in their early phase to soil depletion and erosion. There was a rapid increase in the amount of land under production, largely because of mechanization. Between 1901 and 1921, the amount of land in field crops doubled in Manitoba and increased about 25 times in Saskatchewan and about 45 in Alberta.[73] Mechanization and changes in power sources also affected the size of farms by increasing economies of scale. At the same time, they contributed to higher farm costs and, most importantly, made these costs inflexible by increasing the cash needed for farming. All of these factors in turn had a profound impact on shaping prairie society—declining rural population changed the destiny of many small towns and rural districts, and contributed to the increase in the growth of large cities.[74]

For manufacturers on the Prairies, increasing mechanization and advances in sources of power created numerous opportunities. They not only resulted in the use of increasing numbers and types of equipment in farming, but also made technological innovation ongoing and systematic. Much of the increasing demand was met by the full line companies, which were able to dominate by setting the pace of technological change in sources of power. Nonetheless, manufacturers on the Prairies still saw increased opportunities, and some, like Versatile, were able to compete in the manufacture of large equipment such as tractors. Others found their place in the manufacture of tractor attachments and related equipment. All of this led to growth in implement manufacturing on the Prairies, but this manufacturing was never separated from the wider issues that prairie economic and agricultural development faced: distance from markets, variable commodity prices, and increasingly high capital requirements for farming. All of these issues had an impact on the industry and the products that the market demanded.

Clearing and Tilling the Land

The earliest farm equipment technology used in prairie farming was imported without modification. Prairie farmers, however, soon found that the region's soils and climate demanded equipment that better met their needs. With average annual precipitation under 19 inches per year, and less in many parts of the region, the Prairies are a semi-arid region. Deep frost action in the winter adds to the special conditions. These factors influenced tillage on the Prairies, the history of which falls into two major periods defined by the environmental crisis of the late 1920s and 1930s. In the first period, prairie manufacturers were relatively marginal players in producing tillage equipment, while after the 1930s, the industry developed in the context of significant innovation in dry land farming. It was here that prairie manufacturers found their greatest success.[1]

Developments in tillage equipment responded to theories and findings about tillage practices on the Prairies, as well as the technological framework inherited from the nineteenth century. Before the 1830s, cast iron shares and moldboards were in use, but it was not until the late 1830s that they were manufactured from a single piece of rolled steel. In 1868, James Oliver hardened the wearing parts of a plow bottom without making them too brittle, perfecting this chilling process in the early 1870s. These blades were durable and "scoured" well, that is, the sod did not stick to them and turned over cleanly. Another early innovator was the John Deere Plow Company, which by 1849 was manufacturing 10,000 plows per year. Some of these were riding plows supported by wheels, and some, for larger applications, had several plow bottoms (or blades) on a single machine.[2]

Tilling the land in Manitoba in 1883. On the left is a walking moldboard plow, in the centre a disc harrow, and on the right a single bottom horse drawn moldboard plow. (Ernest Brown Collection, B84, PAA)

Breaking and Clearing the Land

Spring was the time to break virgin prairie sod. In the late nineteenth century, sod was often broken with a single-furrow walking plow, and the coming of the steam traction engine meant this could be done much more quickly. In some cases, the land was plowed two or three inches deep and then replowed to about four inches later in the summer or in the next year following harvest of the first crop. This replowing was called backsetting. Alternatively, the land was immediately plowed four or five inches deep, turning the sod over completely so the vegetation would rot more quickly. In either case, the land was disced and harrowed the same year so a crop could be sown the following spring. The settlers' usual shortage of capital required that they get land into production as quickly as possible.[3] While this was usually done with moldboard plows that were mass produced by full line companies, at least one plow intended specifically for breaking sod was manufactured on the Prairies. A breaking plow, perhaps among the first agricultural equipment to be manufactured on the Prairies, was made by the Vulcan Iron Works, Winnipeg, in 1884. Designed to be drawn by a steam traction engine, it was apparently tested at Regina, although no other information is available about its design or success.[4]

Breaking land covered with trees or brush in the northern and central parts of all three provinces was more complicated than breaking grasslands. Clearing began by removing trees and piling and burning them. The land between the piles was broken immediately and put into crop. The rest of the land was broken after the remains of the burnt trees were re-piled and burnt again. Roots were then picked from the soil and burnt. Until about World War II, a lack of equipment and money meant that most of this work was done by hand. Dynamite was often used to split large stumps and loosen the soil attached to the roots. They could then be pulled more easily with a stump puller.[5] Stump pullers were often home-made from chains and a grab hook attached to a long pole. The stumps were levered out of the soil either by working the pole back and forth by hand or, in some versions, with horses.[6] More complicated, but likely more effective, were commercially produced stump pullers, such as those manufactured by foundries like the Western Foundry and Machine Co. Ltd., Edmonton, in 1919.

Brush land was somewhat more easily cleared than land with large trees because mechanical brush cutters could be used. They were available before World War I and saw widespread use. While there were different types, most were horse drawn. The commonest had a V-shaped blade which sheared off willows, saplings, bushes, and small trees. They were often made by blacksmiths, although manufacturers situated in areas with heavy brush cover usually also met demand in their immediate locale. The Eagle Manufacturing Co.'s brush cutter, Dauphin, Manitoba; the Roose brush cutter manufactured in Wetaskiwin, Alberta; and the Steven's brush cutter, manufactured in Didsbury, Alberta, are examples of such products.[7] While this equipment reduced the labour needed to clear the land, apparently none of it was entirely effective. As one Saskatchewan homesteader observed in 1933, the best way to clear land was to supply axes, cables, and chains

These stump pullers were manufactured by the Western Foundry & Machine Co. Ltd., Edmonton, in 1919. (NC–6–4292, GAI)

to "a couple of good big brawny men with just enough brains to enable them to work intelligently."[8]

Once the brush was cleared, the land was broken. While moldboard plows could be used to break grasslands, they were largely ineffective in breaking heavy land full of roots. Heavy soils often found in wooded areas stuck to moldboard plows, and these problems were compounded because the plow often became jammed in the roots, especially those of willows. And, it was found that most moldboards could not turn such a sod over, but twisted it or left it standing on edge. While many heavy breaking plows were imported to the Prairies, these problems also brought about much local experimentation, and a number of breaking plows were devised by farmers, blacksmiths, and others. Some were commercially produced; among these was a plow produced by the Camrose Plow Co. It was locally designed by Oscar Hough and was sold throughout the Prairies.

Frank Van Slyke, a blacksmith from Iowa who moved to the Red Deer area in 1905, also patented a plow in 1910. It featured a projecting fin and rods and a sharp-angled coulter which flipped the sod over and forced it to lay flat. The same year, he established the Van Slyke Plow Co. The company could not find sufficient capital and ceased operation by 1920, but manufacturing was then picked up by an Edmonton iron-working shop. Van Slyke's plow continued in production until the late 1920s, when demand for breaking plows was falling off because of a declining amount of uncleared land in central Alberta. Yet it had been a popular implement: by the 1920s, over 1,800 were in use in Alberta alone, where it could be advertised without exaggeration as "the famous Van Slyke plow." It was a significant early attempt to create tillage equipment for prairie conditions, and the pride of Red Deer in its inventor was demonstrated by the inclusion of the plow on the town's coat of arms.[9]

With the exception of the tractor-mounted blades that came into common use, the practices established by 1920 for land clearing remained common until World War II. By the early 1940s, however, the general approach to land clearing was changing. Higher farm commodity prices meant that farmers could afford to hire power machinery to clear their land. These clearing outfits used diesel Caterpillar tractors equipped with steel-cutting blades to shear off trees and brush, while bulldozers and mechanical pilers heaped them into piles for burning. By this point, picking roots was the remaining manual task, although mechanization, in the form of mechanically powered rakes that gathered up roots, was making inroads here as well. All of these developments speeded up the process of getting land into production. The land was then broken with breaking plows or discs.[10]

The Van Slyke Plow, developed near Red Deer, Alberta, was typical of the plows devised on the Prairies for breaking heavy soil that contained tree roots. This photo shows the plow in 1926 by which time it was manufactured by the Edmonton Iron Works. (ND–3–3183, GAI)

Two firms typified these postwar changes. In 1954, Fred and Margaret Kellough moved from Edmonton to Camrose, Alberta, to set up a construction company. Kellough also carried out custom land breaking with a Caterpillar tractor and a moldboard plow. In 1958, they moved to Stettler where Kellough began importing breaking plows and discers from the United States. Discs were preferable to moldboard plows because they worked the soil better, but the American equipment was not strong enough. The manufacturer refused Kellough's requests to make the equipment heavier, forcing him to rebuild them for western Canadian conditions. Reasoning that if he had to rebuild them, he might as well manufacture them, Kellough began production on a custom basis. By 1973, he was manufacturing full time, and by 1977, demand had increased sufficiently that he incorporated Kello-Bilt Industries Ltd. By the mid 1980s, the company was producing 12 models of breaking discers as well as other heavy tillage equipment. This equipment was used to break land for agricultural purposes as well as resource development projects.[11]

The story of Dika Industries Ltd., another Alberta manufacturer, is a similar one. In the late 1940s, Mike Dika had a small welding business in Rycroft. In the mid 1950s, he and his brother went into road building and logging. By 1963, lumber prices had fallen and they turned to custom land breaking. It was here that Mike Dika's inventive abilities found scope. He not only built equipment on a custom basis but also designed a breaking plow, as well as some brush clearing equipment. The company's most significant production included brush breakers, tree fallers, and root pickers. These root pickers were among the first equipment devised to mechanize this back-breaking task, and it operated by sweeping roots into windrows.[12]

Even after the land was cleared and cultivation had begun, there was still an ongoing problem with rocks in some parts of the Prairies. This would not have been such a problem if rocks had to be picked only once, but the deep frost of the Prairies kept forcing more of them to the surface. Before World War II, the solution was simple and laborious: smaller rocks were picked by hand while large ones were pulled out with a chain. While newspapers sometimes carried instructions on how to build home-made mechanical rock pickers, it was not until the late 1940s that mechanical pickers were developed with any success.[13] Mechanized rock picking, and the manufacturing that it stimulated, was made possible by changes in the infrastructure of prairie life. Peter Anderson of Southey, Saskatchewan, who designed the first effective rock picker, noted that before the late 1940s such manufacturing was limited because "until welding was perfected it was impossible to build any kind of rock picker strong enough" for prairie needs.[14]

Anderson had established Anderson Construction Ltd. in Southey in 1948 and was primarily involved in machine shop work. The Southey area had rocky land, and picking rocks was one of the burdens of life in the district. Anderson's rock picker thus directly responded to these local needs. His machine established a basic principle that was subsequently used on many rock pickers. It was operated by the tractor power-take-off and used steel-toothed reels or beaters that dragged in the ground and picked up small and large rocks alike. As the reels rotated, the rocks were swept over a grill (so the soil could fall back onto the ground) and were dumped into a steel box behind the reels. The box held about 3,000 pounds. The rocks could be dumped into a pile hydraulically or, with a special attachment, lifted and dumped into a wagon. Anderson manufactured these rock pickers at his shop in Southey and had produced 40 machines by 1953. Unable to meet demand, he turned manufacturing over to a Montana firm which manufactured them on a royalty basis. In the 1970s, Leon's Manufacturing of Yorkton purchased the rights to Anderson's rock picker. Leon's was already manufacturing one type of rock picker, and this purchase expanded its line and diversified its already extensive product line of dozer blades and other soil-moving equipment.[15]

Peter Anderson was the first of a long line of manufacturers producing rock pickers. While they were technologically significant, his machines were expensive and so did not stimulate widespread adoption of rock pickers. Instead, this occurred in the early 1960s with the involvement of other manufacturers who were able to produce less expensive equipment. Most of these firms were located in Saskatchewan, and their success lay in producing affordable equipment that met varying needs in many parts of the world. One such manufacturer was Degelman Industries. In 1963, Wilfred Degelman designed a rock picker on his farm near Raymore. Like the Anderson machine, it employed the rotary principle, although Degelman modified it in some respects. He built 10 machines on his farm that year, and by 1965, his volume of production justified building a plant in Regina. He soon added other equipment to his line, especially bulldozer blades. During the early 1970s, Degelman Industries manufactured blades for John Deere, which not only led to an expansion of the business but gave it distribution through John Deere dealers in Canada and the United States. While the Deere contract represented 50 percent of the firm's production by 1973, rock-picking equipment was still Degelman Industries' central product line. In the same year, the company introduced a rock windrower to the market. This machine speeded up the task of rock picking by dragging rocks into a row where they could be gathered up by a rock picker.[16]

Two other business contemporaries of Wilfred Degelman were John Schulte of Schulte Welding and Machine Ltd., Englefeld, Saskatchewan,

Among early manufacturers of rock pickers was Rock-O-Matic of Vonda, Saskatchewan. The machine was first designed by Rosaire Bussiere in about 1960. Using some of the same principles as on the 1960 model, this one was manufactured in the 1980s. (PAMI/Alberta Agriculture Photograph)

Machines to gather rocks into a windrow where they could be more easily picked up with a rock picker were another prairie development. This one was manufactured by Degelman Industries, Regina, in the mid 1980s. (Degelman Industries Ltd. Photograph)

and Rosaire Bussiere of Vonda, Saskatchewan. Schulte started to manufacture brush-cutting equipment in 1942, but the most important development came in 1948 with his production of a snow blower. This proved to be a popular line, and in 1958, he also began production of a rock picker. Unlike the Anderson design, it used an elevator-type conveyor to carry the rocks into the bin. The 1970s saw a name change to Schulte Industries Ltd. and a number of new products including a reel-type rock picker and a rock windrower, as well as new models of snow blowers and earth scrapers. This expanding product line was further demonstrated in the mid 1980s with the company's production of a rotary brush and grass cutter for clearing small trees and bush on rough and uneven ground.[17]

The manufacture of Rosaire Bussiere's rock picker began on a small scale as well. Bussiere was a farmer who built his first rock picker in 1961. The same year, he set up a company called Rock-O-Matic in Vonda to produce them. They employed a rotary principle, and like other rock pickers, they had special features that made them unique. For example, the bin had a grated bottom so that the soil could fall out. In 1976, the company developed a combined rock picker and windrower, the first of its kind in North America.[18] In 1978, the company was sold to Smith-Roles, and Bussiere's son Ray, who had worked with his father, continued to manage the plant at Vonda. When Smith-Roles went into receivership in the late 1980s, Ray Bussiere bought the Vonda plant, restoring the family's ownership, and called the firm High-Line Mfg. Inc. Until this point, the plant had produced only rock pickers in a wide variety of models. It was obvious to Bussiere, however, that diversification was essential for long term success. Thus, in 1988 he purchased

Laurier Manufacturing of Laurier, Manitoba, an established manufacturer of round bale equipment and packer/harrow bars. This provided a solid expansion of product lines.[19]

The experience of Degelman, Schulte, and Bussiere indicated that while production of rock pickers could provide a sound business base, they were not enough to sustain expansion. Although there was a tendency for farmers in the 1960s to share a single machine, this seems to have declined in following years. Sales of rock pickers were steady, although they were highly regionalized. Since rocks were a problem only in certain parts of the Prairies, such as north and east of Regina and near Saskatoon, it was no accident that firms specializing in rock pickers were located in these areas. It was a market dominated by short line manufacturers because no full line companies would produce rock-picking equipment given the small market. Export markets, especially in the United States, were important and most of these manufacturers, at one point or another, maintained distribution depots in the U.S. While there were some American competitors, Canadian short line companies were able to gain significant penetration of the American market.[20]

Summerfallowing

Mennonites and other settlers in Manitoba were using summerfallowing in the late nineteenth century. This practice involved leaving the field fallow and tilling it regularly to kill weeds, which used the moisture in the soil. While this early example of summerfallowing was notable, precedents in Dakota and Nebraska were of more direct significance for the prairie west.[21] But the practice also developed from prairie conditions and experience. Its adoption—largely by accident it seems—in south-east Saskatchewan in the mid 1880s also served as a local demonstration of its value. Theorists like Angus McKay, superintendent of the experimental farm at Indian Head, also added their support by arguing that summerfallowing increased production. By 1900, these views had coalesced into a general theory of dry land farming that emphasized deep plowing, keeping the surface soil loose and level and the subsoil compact, continual cultivation until freeze-up, adding organic matter to the soil, summerfallowing in dry areas, and varied crop rotations.[22]

Summerfallowing was a costly system: the continual tillage created high labour demands, and, of course, there was no production for one year on a summerfallowed field. It remained a subject of perennial debate on the Prairies, with proponents of continuous cropping arguing that summerfallowing was unnecessary, especially in areas with adequate moisture. Nevertheless, it was generally thought to be required for retaining moisture and preventing crop failure, and even though a year's production was lost, increased production in other years was said to offset this loss.[23] By 1900, summerfallowing had garnered further support when it became generally accepted that a dust mulch, a three- or four-inch layer of pulverized soil, would help conserve moisture. This approach was promoted by a North Dakota farmer and was soon accepted everywhere in the dry-belt areas of the

Prairies. A finely tilled black field was taken to be the mark of a good farmer, a source of pride and accomplishment that came from the ideal of the European garden, not the prairie farm.[24]

This cultivation was achieved by primary and secondary tillage operations. Primary tillage was most often performed with a moldboard plow which turned the soil over completely, burying weeds and stubble. The exposed soil became loose and dry from sun, air, and frost. The earliest moldboard plows had been single-bottom walking plows, but between about 1910 and 1920, three-bottom plows drawn by horses became typical. Depth was adjusted with hand-operated levers. This often required two workers: one driving the team or the tractor and another to work the levers. Different types of bottoms were available for different soil types or tasks. For clay soils that stuck to the moldboard, disc plows—which also had a lighter draught—had been introduced by about 1910. Disc plows were also popular in areas heavily infested with couch grass. They were used to open the soil so that the root system of the grass could be pulled out with a spring-toothed harrow or cultivator. However, because they did "not cut grass roots nor cover weeds, nor pulverize the soil as well as a mouldboard plow," moldboard plows retained much of their overall attraction.[25]

In keeping with the overall objective of summerfallowing, the primary aim of secondary tillage operations was to prevent moisture loss. Packers conserved moisture by pressing down and levelling the newly plowed soil, and were often attached to plows to combine field operations.[26] By 1919, tests showed little advantage to packing in higher-moisture and heavier-soil areas such as those around Brandon. Although there were exceptions, packers were therefore most common in drier areas with lighter soils in southern Saskatchewan and Alberta.[27] While a variety of packers were available, they generally fell into the two broad categories of surface and subsurface types. Surface types pulverized the soil and eliminated clods of earth. They were often also used for packing down seed after seeding. Subsurface packers had thin, narrow wheels which pressed down the subsoil.[28] Some of these packers were rigid-frame machines, while others were flexible so they could follow the surface of the ground. All could be towed by a plow or used independently. The objective was to make a field level but not too smooth as this would encourage evaporation and, ultimately, soil drifting. While all packer types were used widely on the Prairies, Seager Wheeler, the prize-winning Saskatchewan wheat grower, strongly advocated the use of packers that only pressed down the soil, rather than those that churned it. This was especially recommended when a packer was used before seeding since the resulting aeration of the soil dried it out.[29]

In addition to the packer, drag harrows were used to level the field, to kill small weeds, to create a dust mulch, and, towed behind seeding equipment, to ensure coverage of the seed. Level fields were not just a fashion. Seager Wheeler contended that binders and seed drills worked best on level fields (in part because they were ground driven); indeed "every implement used on such [level] land will operate more satisfactorily." There were several types of drag harrows commonly

These gasoline tractors are towing large moldboard plows. The one on the left has seven bottoms, the one on the right is towing a packer behind the plow. Probably from southern Alberta, not dated. (Archives Collection, A1315, PAA)

used, including spring tooth, chain, and spike tooth harrows. The first two were specialized harrows used mainly to control couch grass. The spring tooth harrow brought the grass roots to the surface so that they could be gathered up by a chain harrow and then burnt. Spike tooth harrows were used to clear small weeds from fields.[30]

This shallow tillage could also be achieved with disc harrows. Although they had a tendency to plug on heavy soils, they were among the earliest shallow tillage implements used on the Prairies. They were used both on plowed land and for surface cultivation of stubble fields.[31] Available with full or cutaway discs, the disc harrow's rolling action worked in considerable amounts of stubble or pulverized the soil on already plowed land. They also were towed behind a plow to give added tillage, or behind a binder to stimulate growth of weeds which would then be killed by the frost before they had set seed. Seager Wheeler believed that disc harrows were "indispensable implements," but he noted their tendency to cause soil drifting when used incorrectly.[32]

By the end of World War I, cultivators were replacing disc harrows and moldboard plows for seed-bed preparation, summerfallowing, and fall cultivation of stubble fields. This development was most pronounced in Manitoba. As observed in 1920, "farmers are not plowing their summer fallows" with moldboards because they found the cultivator better for killing weeds and conserving moisture.[33] Within a decade, cultivators were also in wide use in those parts of the Prairies where disc harrows caused soil drifting. While cultivators tended to plug in fields with heavy stubble, they were more adaptable than disc harrows because they could be fitted with various interchangeable cutting points. The most common of these on the Prairies by 1920 were duckfoot blades which cut weeds below the surface. Diamond point types were also used for digging up hard ground or creeping rooted weeds like quack grass, but they were not widely used because too much power was needed to drag them through the soil.[34] Thus, cultivators were most often equipped with duckfoot blades and worked best on light soils that had been levelled with a packer or harrows. In 1919, Seager Wheeler foretold that the cultivator was "destined to play a more important part on the farm in the future when it is better known. It can be used very effectively without destroying the texture of the soil; in fact, it will improve it."[35]

This farmer is using a disc harrow to produce a finely worked surface on his summerfallowed field in southern Alberta in about 1920. (Harry Pollard Collection, P624, PAA)

Seager Wheeler's concern about soil texture showed foresight. In the dry years of 1917 and 1918, soil drifting was noticed in Manitoba where even the heavier soils typical of the province were showing signs of deterioration.[36] But the truly major crisis lay only a few years in the future. By the early 1920s, because of poor soil management and dry conditions, soil drifting and soil degradation were a serious problem in parts of southern Saskatchewan and Alberta. Low rainfall in the early 1930s worsened these problems, creating a crisis throughout the dry belt of the Prairies. In some places, soil fertility that had taken thousands of years to accumulate had been destroyed in a single generation.

Manufacturing Equipment for Black Summerfallowing

By the 1850s, the Red River settlers were importing American and British moldboard plows.[37] Mass-produced plowing equipment from Canadian and American manufacturers was also readily available to those who later settled the Prairies. The full line companies therefore dominated the field, and only a few prairie firms attempted to compete in this market. The Ronald Smith Cultivator Co., Winnipeg, advertised a cultivator in 1919, but it seems to have been a short-lived effort.[38] In the mid 1920s, the Lethbridge Iron Works manufactured a disc plow based on designs by Ortus Henry, a Lethbridge-area farmer. Called the Buffalo Disc Plow, it was a tandem-disc machine that cut the soil deeply, but it was in production for only a few years. By the late 1920s, machinery like this was becoming obsolete in dry land areas because of concerns about soil conservation and a growing trend away from deep tillage.[39]

In the era when black summerfallowing was popular, the only manufacturing that took place on the Prairies was in packers and, to a lesser extent, harrows. Packers were important for dry land farming, and regional production had an obvious advantage because packers were relatively inexpensive but were heavy and awkward to ship. Most companies were located in Winnipeg and included the Sub-Surface Packer Co., active at least from 1909 until 1914; the Hamilton Pulverizer Co. which was operating in 1909; and C.W. Anthony who manufactured a smooth roller-type corrugated packer and pulverizer in 1914.[40] A few firms tried their hand elsewhere on the Prairies. The Brandon Machine Works produced a packer from 1908 until 1914, and the W.H. Church Co., Saskatoon, was incorporated in 1909 by a farmer and two Saskatoon businessmen to acquire Canadian rights to a steel-frame reinforced cement land packer. It changed its name to The Western Foundry and Machine Co. in 1911, but unable to attract investors, it was wound up in 1914.[41]

The economic upswing in agriculture during World War I brought a number of other companies onto the scene, but few survived the economic decline following the war. In 1917–18, the Ideal Harrow Co. of Foxwarren, Manitoba, produced a flexible harrow for use behind a gang plow.[42] Between 1914 and 1918, the Christiansen Harrow Co., Winnipeg, produced a variety of harrows, including a rigid-frame spike tooth harrow. The company was innovative—in 1916, it brought out a circular harrow, and a year later, introduced a system in which the harrow folded up against the plow for transport. This was among the first attempts to accommodate the transportation needs created by large farm size. The company also produced land packers. By 1917, it had apparently shifted its efforts into an automatic stooking machine and operated as the Bell Automatic Shocker Company. Nothing was heard of the company after 1918.[43]

Packers nicely demonstrate the problems faced by early farm equipment manufacturers on the Prairies. While they were a logical product to manufacture because of their weight and consequent freight costs, they provided an insufficient base for regional industrial production or development. The full line companies produced them, which pitted well-organized and well-capitalized competition against local firms, and the market for packers did not extend to all of the region. They were used primarily in light soil areas, which had largely been settled by 1914. The greatest subsequent expansion in settlement was in parkland and heavy soil areas, where packing was not practised as widely. Moreover, packers did not wear out, and a farmer would likely have had only one during all of his farming years. So too, if the advice in farm newspapers is any measure, a number of farmers either shared a single packer or made their own from worn out disc harrow blades or even discarded threshing machine cylinders.[44]

These patterns endured until the 1930s when, because of the environmental crisis on the Prairies and the development of new and specialized approaches to tillage, greater opportunities for manufacturing of tillage equipment emerged. The differences between the two periods lay in a combination of factors. Where there was a particular regional requirement, such as for specialized plows to break brush land, a number of prairie firms met these needs. However, on broken land, the tillage practices of black summerfallow were standardized within a relatively small range of equipment which was produced by established Canadian and American firms that adequately fulfilled farmers' needs. Their equipment was seen as universally applicable to the Prairies: all that was needed was a moldboard or disc plow, harrows, possibly a packer, and a duckfoot cultivator.

Responding to the Environment: Changing Tillage Practices

The environmental crisis of the interwar years led to developments on a number of fronts. The federal government became involved through the 1935 *Prairie Farm Rehabilitation Act* (PFRA) which promoted water conservation and soil reclamation. In cases of very serious soil degradation, land was transferred to the crown, seeded with grass (some with drought-resistant varieties developed in the 1930s), and turned into pasture land. Among these projects were those in the Tilley East area in Alberta, at Cadillac in Saskatchewan, and at Melita, Manitoba. Reclamation efforts by PFRA also included educational work through agricultural improvement associations to teach methods for preventing soil drifting. Sometimes in co-operation with the federal government, and in other cases on their own, the provinces also developed programs promoting soil conservation. Saskatchewan, for example, offered short courses in agricultural techniques. As well,

Soil erosion because of wrong tillage practices and equipment became a serious problem throughout the dry land areas of the Prairies. This photo shows erosion in southern Alberta in the 1930s. (Archives Collection, A5659, PAA)

research and experimentation were carried out at universities and research centres throughout the Prairies. It was a time of significant co-operation among farmers, scientists, and government.[45]

A significant change in tillage practices was essential if the crisis of the 1920s and 1930s was to be met. The old methods of working the soil finely and leaving a dust mulch were gradually abandoned. Pioneering developments in strip farming took place in southern Alberta after World War I and were in use there by 1930, although it did not become a common practice elsewhere on the Prairies. Similarly, a new implement, the basin lister, was used. These were built in southern Alberta and parts of Saskatchewan by farmers, by the Swift Current Experimental Farm, and also by full line companies like John Deere and IHC. A lister plow, which left alternate ridges and furrows, was adapted with an attachment behind each spade which created dams along the furrow every few feet to catch rain and run-off.[46]

A more widespread change came with the use of new primary tillage equipment that cultivated more shallowly and left a mulch, or residue, of stubble and weeds (or "trash") on the surface. This helped to retain moisture and prevent soil drifting. Duckfoot cultivators were popular by the late 1920s for weed control since they did not pulverize the soil.[47] But these cultivators tended to clog in heavy trash and were too light for primary tillage. Other solutions were necessary, and among these was the development of the one-way (often called a "tiller-combine"), which evolved out of disc plows and disc harrows.

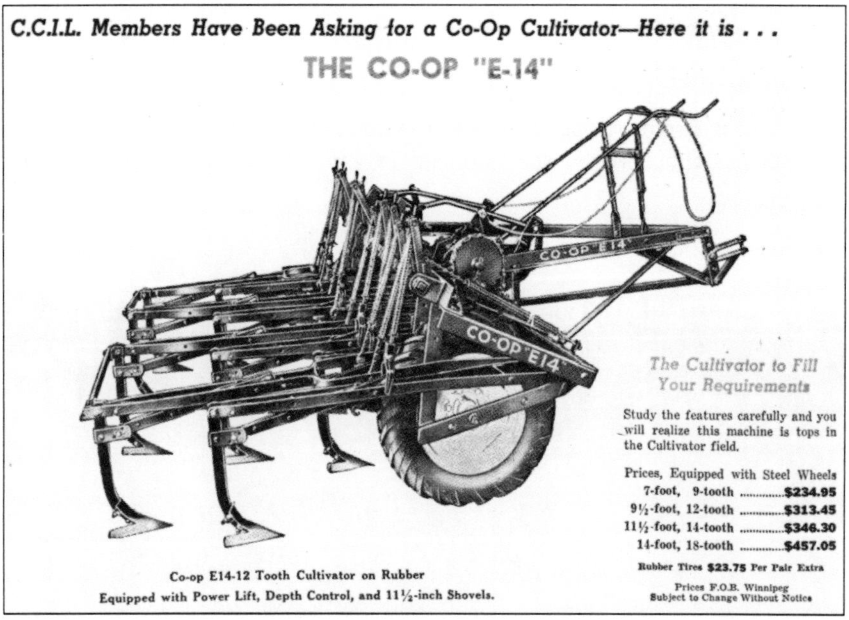

Cultivators with duckfoot blades, such as this one manufactured by CCIL of Winnipeg in 1950, were an early alternative to moldboard plows. (*CCIL Farm Machinery Catalogue 1950*)

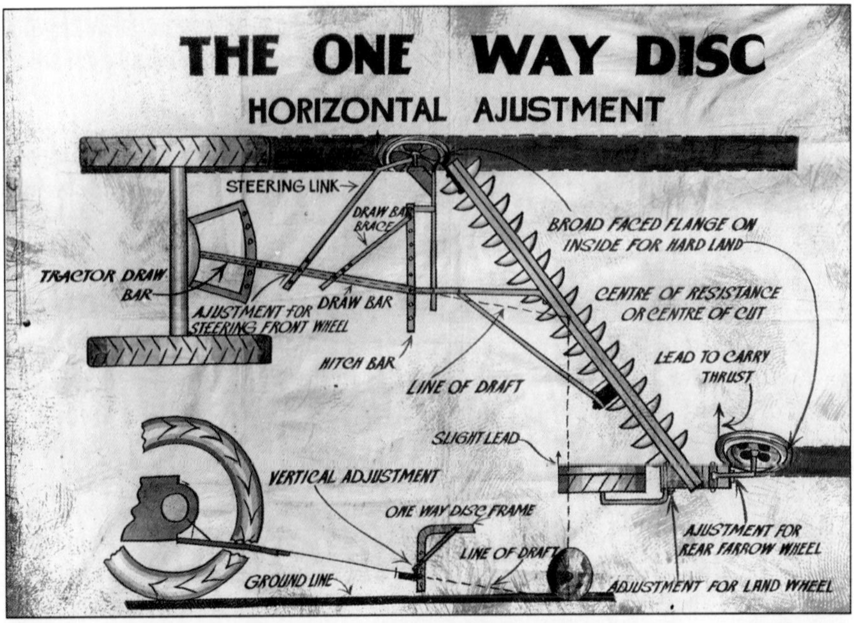

The one-way was a new implement introduced to the Prairies in the 1920s. It was difficult to hitch, and instructions such as these helped to guide farmers in their use. (Public Affairs Bureau Collection, PA1420/1, PAA)

The advantage of the one-way was that it tilled more shallowly than a moldboard or a disc plow and deeper than a disc harrow. It cut the weeds and worked the stubble shallowly into the soil. In the mid 1920s, a United States patent was issued for a "new type of disc plow built with the discs on one axle and moving the soil one way." This patent was later "upset on the grounds that the principle had been in use in Australia for several years," and most full line companies were manufacturing one-ways by 1930. Since the one-way could also be fitted with a seed box to combine tillage with seeding, it served both economic and soil conservation needs. With this machine, "one operation then opened the soil to a controlled depth; moved the layer of soil to the side far enough to anchor the trash, weeds, stubble, and straw to reduce wind erosion; cut off the spring weed growth; and placed the seed in the furrow and covered it for germination."[48]

One-ways were a highly significant development. They were heavy machines that worked the straw into the soil better than other commonly used equipment and so they made a major contribution to the trash-farming method. In the 1930s, this was called "plowless" farming, a significant phrase that showed the magnitude of change in tillage. One-ways often caused soil drifting in arid districts or on fields without trash, but when operated correctly, they generally did not pulverize the soil and left a cover of straw and stubble. Anchored in the top layer of the soil, this "trash" protected the soil from wind and water erosion.[49] Indeed, it was observed at the Swift Current experimental station in 1931 that the one-way "has so thoroughly proven its worth in preventing soil drifting, cutting the cost of summerfallowing and spring preparation of the stubble that it has been adopted generally for these operations in the areas subject to drought and high spring winds."[50]

Yet, while shallower cultivation was now the general rule, this change-over in equipment occurred unevenly across the Prairies. The new methods were adopted more quickly in some places than in others, depending on local conditions and needs. It was not, for example, until the early 1940s that one-ways replaced moldboard plows and disc harrows in west-central Saskatchewan, even though the change had occurred in drier areas further south in Saskatchewan by the beginning of the 1930s. In frontier areas, moldboard plows continued in use in the 1940s, "although there is some interest in one-ways even in these districts." By 1951, moldboard plows were "practically a thing of the past" in western Canada, especially in dry land areas. As Evan Hardy explained to a Welsh firm hoping to import them to the Prairies, they had been replaced with subsurface tillage cultivators and blades as well as one-ways and discers.[51]

Despite their advantages, one-ways were "cantankerous" implements that were difficult to hitch, needed additional weight in heavy soils, and had considerable side draught. Given these problems, the University of Saskatchewan organized field days in the 1930s and early 1940s to teach farmers how to operate them properly. As O.L. Symes of the University's agricultural engineering department noted, Saskatchewan farmers "have always been innovators," and by the mid 1940s, they were modifying the one-ways (which had 22-, 24-, or 26-inch discs) by fitting them with smaller 18- or 20-inch discs that had been used on disc harrows. Smaller discs had shallow concavity and left more trash on the surface, and they also reduced the power needed to pull the machine. The machine's width could then be increased without adding to its draught, in this way speeding up tillage operations. They cultivated more deeply and did not leave ridges as did disc harrows, but they made a shallower cut than one-ways and left a level surface with a reasonable amount of trash anchored in it. And, like the one-way, they could be fitted with seed boxes.[52] These machines became known as discers.

Testifying to the interest and level of experimentation, by the 1940s, many machine shops and blacksmith shops as well as many farmers were building discers. Because of the war, parts were hard to come by and "junk yards were scoured for heavy steel pipe, old plows, one-ways and other machines were dismantled for wheels, shafting, discs, bearings etc. A distinct change was rapidly taking place in Saskatchewan tillage."[53] H. A. Lewis, a farmer from Gray, Saskatchewan, who also taught at the University of Saskatchewan Agricultural Engineering Department, recalled that the department "had countless men come to see us with part plans, partial ideas and in some cases just plain problems to which they hoped we might have an answer. In many instances we learned more from them than they learned from us." By then, there were many home-made discers operating on numerous Saskatchewan farms.[54]

By the mid 1940s, most full line companies had staff visiting Saskatchewan to evaluate these machines, but CCIL was the first to bring one out. In 1946, it asked

H.A. Lewis to develop a prototype discer applying the best features he had observed and tested. The machine was tested on his farm, and the first machines came off the CCIL floor in 1947. It was the "first large factory production of the discer." The novelty of the machine was shown, as it had been with the one-way, by the fact that there was no common name for it. Some called it a one-way disc harrow, which was logical given its technological pedigree, but it was too cumbersome for popular use. CCIL copyrighted the term "Disker" and other manufacturers used their brand names. Although some of the farmer-designed discers (such as ones devised by W.J. [Wes] Kirby and A.A. [Al] McKinnon) had featured a seeding attachment that had proven effective in field work, CCIL's did not have a seeding attachment. Competing models had this feature, and in 1948, CCIL added seed boxes to its discers. The entry of the full line manufacturers in the field demonstrated the general appeal of this equipment and it soon spelled the end of blacksmith and farmer production of discers. They were being manufactured by Cockshutt in 1947 and by IHC and Massey-Harris in 1948, and they continued to be produced on the Prairies by CCIL, as well as by Killbery which manufactured them on contract to Oliver.[55]

The discer was a Saskatchewan development and was first manufactured by CCIL. It overcame many of the problems of one-ways and was suited to trash-cover farming. The 1950 CCIL "Disker" came with an attached seed box. (*CCIL Farm Machinery Catalogue 1950*)

The discer was a significant prairie development. It helped in some districts to prevent soil erosion and it permitted combined tillage and seeding. Its larger size—made possible by its lower draught—also confirmed and met the demands for larger equipment needed on farms of steadily increasing size. These principles ensured the continued popularity of disc tillage, and with modifications, discers continued to be produced into the 1980s. As well, tandem discs (the descendants of the disc plow and harrow) found increased use in the late 1960s and were manufactured by various firms.[56]

Subsurface Tillage: A Radical Innovation

Discers were not the only innovation in heavy tillage equipment that took place on the Prairies. Some of the most famous and important of these developments were those of Charles S. Noble of Nobleford, Alberta. Noble farmed a large spread in southern Alberta—in 1919, he had over 35,000 acres of land. In 1922, he went

Another type of disc implement was a tandem disc, such as this one manufactured by Ezee-On Manufacturing of Vegreville, Alberta. (Public Affairs Bureau Collection, PA6478, PAA)

bankrupt, but he started again and had another large farm by 1935. He recognized the need for new types of equipment if grain farming was to be effectively carried out in arid areas such as southern Alberta. While the one-way and the discer left some trash on the surface, both implements, by turning the soil over, sacrificed moisture. Noble's approach was more radical. When he was in California in 1935, he observed a sugar beet digger at work. It had a straight sharp blade that worked below the beets and lifted them to the surface. He reasoned that a similar blade that cut under the stubble to kill weeds yet left the stubble standing would prevent soil erosion and trap snow in winter and decrease spring run-off.[57]

Based on this principle, Noble devised a system that used a straight sharp blade about nine feet long fastened to a frame by heavy supports. This was the first Noble blade, although Noble later adopted the use of wide V-shaped blades for his subsurface tiller. He saw that if the weeds were cut below the surface without turning the soil, the weeds and stubble left anchored in the soil would provide protection from wind erosion and evaporation. This was especially valuable for fall cultivation of stubble because the standing stubble caught the snow. Noble tested the blade on his own farm and on others in the neighbourhood. He also arranged for its testing at prairie experimental farms. Because these subsurface tillers were of little use in preparing a field for seeding, their adoption initially was "temporarily sidetracked by the general adoption of the more adaptable one-way disc." It was also found that this equipment worked best in dry areas. Where there was higher moisture, the weeds cut off by the blade and left standing simply took root again.[58]

Noble's work was important and innovative, as recognized by the awards and honours he received: an OBE, an honourary doctorate from the University of Alberta, and, among others, election to Alberta's Agricultural Hall of Fame. But the importance of his inventions cannot be fully appreciated unless they are considered in light of where he lived and worked. The Lethbridge area was subject to severe soil drifting as early as 1915, and by the early 1920s, farmers in the area, as well as the Lethbridge Experimental Farm, had made significant innovations in cropping and tillage practices. So important were these developments that they formed the basis for PFRA's campaign to control soil drifting on the Prairies. By the mid 1930s as well, southern Alberta farmers had adopted trash cover to control drifting. They found the available implements of little use for such tillage; cultivators worked poorly in heavy stubble, and one-ways worked the land too much. Certainly, they recognized the need for new implements that could cut weeds yet leave the stubble standing.[59]

In 1936, C.S. Noble of southern Alberta invented a cultivator that tilled below the surface. Using a straight blade, it reduced surface disturbance and promoted soil conservation. This one was restored at Nobleford, Alberta, in 1969. (NA–4884–29, GAI)

Similar concerns existed everywhere on the Prairies and in some cases had been expressed from an early date. In 1909 for example, R.D. Smith of Franklin, Manitoba, patented a blade weeder, and in 1914, the Guarantee Iron Works, Winnipeg, produced its Murks Thistle Eradicator, a six-foot V-shaped knife that operated as deep as 12 inches below the surface. Another example of this early interest in subsurface tillage was the Great Cyclone Weeder, designed and manufactured by G.L. Dodds in Winnipeg in about 1913.[60] But, in spite of these innovations, few prairie farmers yet felt the need for such equipment. The exception was in southern Alberta where farmers, as early as 1919, were using subsurface tillage equipment to cut weeds without breaking up the soil surface.[61] Interest in this principle grew in the 1930s when several people began work on subsurface cultivators. Among them were J.B. Turner and Otto Wobick of Barons. In 1935, they "started with an old steam plough and attached cultivator sweeps, made of ploughshares, to the beams. Many changes have been made to the original model, commonly known as a 'Paul Bunyan' weeder. Models since have featured cultivators with from one to five duckfeet."[62]

Other people were also working on similar cultivators. Victor Erdman, who was born in the Barons area, devised a subsurface cultivator with three large, wide, V-shaped blades of the sort Noble later adopted for his equipment. By 1940, three general types of subsurface blade implements had been in use in southern Alberta for several years. They included those using the straight blade developed by Noble, those using a single wide duckfoot blade developed in the Monarch area, and those with a large gang of three or four wide V-shaped duckfoot blades devised by Turner and Wobick as well as by Victor Erdman. Because the V-shaped blades were best for light soils and cut more shallowly, they became the most popular. All of these cultivators could penetrate hard ground, and all had large blades positioned so they could "lift the soil and disturb it sufficiently to kill small weeds, at the same time leaving the stubble and other trash undisturbed on the surface."[63]

One type of subsurface cultivator that became common used wide blades. This one was developed by Otto Wobick of Barons, Alberta, in 1935. In this 1965 photograph, Wobick poses beside his cultivator. (The Archives, Sir Arthur Galt Museum, Lethbridge, P19760204078)

Although these machines were often used only by the inventor and perhaps a few neighbours, commercial production did follow in some cases. The only full line company to produce subsurface cultivators was John Deere, but since its machines were too light to find favour in southern Alberta, most production was local. Victor Erdman had a welder (powered by a generator) on his farm, and he began commercial production of his three-blade cultivator in the early 1940s under the name Victory Manufacturing Co. These machines found a local market, and in 1949–50, he built 100 machines. In 1950, Erdman moved to Lethbridge and changed the company name to Victory Equipment Ltd. He entered a partnership with Lawrence Edwards (later to establish a company making rod weeders) and Wally Strom (brother of a future Premier of Alberta). Erdman bought out his partners in the late 1960s and then sold his company to local interests in 1977.[64]

Other subsurface cultivator manufacturers in Alberta included Johnson Tillivator, formed by Lloyd Johnson, a Barons area farmer who had designed a three-sweep subsurface 11-foot cultivator. George Gareau, who owned an iron shop in Barons and later moved to Vulcan, was also among the many others who produced such cultivators.[65] The most important of these manufacturers, however, was C.S. Noble. He built his cultivators on his farm in 1937–38, and began selling them. While absorbing immense amounts of time and cash, the cultivator was still a secondary concern to the farm. Noble was often in precarious financial shape, and as he admitted in 1938, he could not "allow this cultivator business to interfere with our farming."[66] It soon moved beyond secondary status, however, and in 1943, he established a factory in Nobleford. By 1946, he had produced about 2,100 blade cultivators, most of which were sold in the arid sections of Alberta, Saskatchewan, and the United States.[67] Noble continued to experiment and test his machinery, and tried to meet changing conditions. In the early 1950s, for example, Noble Cultivators began equipping its machines with hydraulic lifts to keep pace with general technological change.

Charles Noble died in 1957, but his children continued the business. By 1972, it was producing blade cultivators, harrows, and a rod weeder attachment, as well as seed drills. In 1976, the company expanded its plant in Nobleford by adding 60,000 square feet of space, and by this point, it was the largest farm implement manufacturer in Alberta. It employed 200 workers and enjoyed regional and international sales.[68] In the late 1970s, the company was purchased by Versatile and became a subsidiary called Versatile Noble. Reflecting increased farm size and speed of operations, Versatile Noble's blade cultivator in the early 1980s was equipped with nine bottoms, totalling 50 feet in width. The wings of the cultivator hydraulically lifted for transport. In 1987, Versatile sold the Noble plant, and the Noble lines were taken over by New Noble Services. New Noble added them to its existing lines which mainly included the Victory lines (including Victor Erdman's V-shaped blade). New Noble marketed the Noble lines through Vicon Manufacturing Inc., while the Victory line was sold through New Noble's distributors.[69]

Noble's work was significant because it took the idea of trash farming to the furthest possible point. In addition, it demonstrated one important aspect of the process of invention and technological change. The Noble blade was developed within a specific environmental and economic context and met practical and immediate needs. However, while the principles of Noble's cultivator were accepted, his machines encountered initial resistance because of their appearance. While most farmers had "high praise" for the blade cultivator's work, Noble was told in 1938 that "the skeleton like appearance of the implement seems to detract from the high price asked," despite "the intrinsic value of the work performed." Fashion and traditional styles were powerful forces in defining value for farmers.[70]

The development of subsurface tillage equipment in southern Alberta took place within an environment that stimulated invention and experimentation by numerous farmers and scientists. It was an Alberta parallel to the development of the discer in Saskatchewan. Public and private sectors worked together to learn from each other's observations and theories. Materials and equipment, which were often difficult to obtain because of cost or supply, were often shared among local manufacturers. While Noble was always concerned about protecting his patents, and undertook action to do so in a number of cases, he "never considered local manufacturers competitors, but friends and sold most of them steel at lower prices than they could buy elsewhere."[71]

Heavy Duty Cultivators

The desire to leave a trash cover was met by another new piece of equipment, the heavy duty cultivator (also known as the chisel plow), which became available immediately after World War II. This machine featured heavy shanks with various types of flat shovels. It was "a valuable machine to maintain even and uniform depths in soils with both hard and soft spots and at practically any tractor speed." It had good clearance, and so did not plug as did the older types of cultivators, and it left a large amount of trash on the surface.[72] Essentially an American development, there were some early attempts to use it on the Prairies. In an effort to cultivate through the heavy trash cover that he had left to prevent soil drifting, a Manitoba farmer reported in 1938 that he had removed the moldboards from a gang plow, leaving the shanks as the cutting points. He called it a "cultivator plow" and praised its ability to clear trash better than the duckfoot cultivator.[73]

Heavy duty cultivators balanced the change to trash cover tillage with the need for increased speed of farming on the Prairies, thereby suiting the regional nature of prairie short line manufacturers. While full line firms were important manufacturers of heavy duty cultivators by the late 1960s, about 40 percent of prairie needs were supplied by prairie firms.[74] And in keeping with this regional market, there was ongoing redesign by prairie firms to meet changing needs. CCIL brought out a heavy duty cultivator in 1953 called the "plower." It cultivated at a uniform depth but left trash on the surface.[75] Another early manufacturer of heavy duty cultivators was Brandt Electric Ltd. of Regina. The firm was established in 1938 by A.P. Brandt to do electrical work, but by 1946, it was manufacturing grain augers. It soon made other products as well, including a heavy duty cultivator. In 1951, it was producing four sizes of cultivators up to 14-feet wide, all equipped with hydraulic controls. Since some early heavy duty cultivators had a tendency to cultivate unevenly in depth, Brandt dealt with this problem by controlling depth of cultivation with an adjustable collar on the hydraulic ram. Production of these cultivators continued, and by the early 1960s, the company was also producing grain augers, rock pickers, and land-levelling equipment.[76]

By the mid 1960s, even larger heavy duty cultivators were being used. However, as their size increased, so did problems of ensuring even tillage depth and reducing draught. In the early 1960s, Roll-O-Matic Tiller Ltd. of Regina manufactured a heavy duty cultivator with a unique type of shank holder and trip mechanism that reduced draught requirements.[77] Other problems were also being addressed. Peter Anderson of Southey, who had developed the rotary rock picker, devised a shank release system to prevent breakage of the cultivator frame or the shanks in rocky land. He also set a significant precedent with his construction of a cultivator in two sections hinged at the centre. This allowed it to follow rolling land to maintain consistent depth. His largest cultivator in the early 1960s was 21 feet, but in 1962, he was planning a 69-foot model with a flexible frame.[78] As the size of both equipment and farms increased further, transportation problems became more pressing. Anderson's cultivator met this need by having the sides of the cultivator lift up for transport.[79] By the 1980s, heavy duty cultivators had become the most widely used tillage implements on the Prairies. They could be used for tillage as well as air seeding, and there was a wide variety of shovels for specialized use.[80]

By the early 1950s, heavy duty cultivators had become an alternative to disc implements. With their higher clearance, they left a layer of trash on the surface and promoted soil conservation. This one was designed and manufactured by Anderson Construction of Southey, Saskatchewan, in the early 1960s. (Agricultural Machinery Administration, Province of Saskatchewan, *Report on Test of Anderson Heavy Duty Cultivator,* 1963)

These developments showed that increasing the size of equipment was not always simply a matter of making it bigger. The mechanical principles often had to be reworked to accommodate the increased size; in effect, a new design was necessary. This was demonstrated by the work of Olaf Friggstad, a farmer near Frontier, Saskatchewan, who began manufacturing cultivators, rock pickers, and bale trailers on his farm in 1969–70. Friggstad's motives were supremely practical: he was "a bit of a handyman and workshop tinkerer" who had five sons to provide for. He also saw a need for larger and more rugged cultivators than those that were available,[81] and in 1969, he put together a 40-foot unit of heavy construction. Given

By the late 1970s, heavy duty cultivators were in use on most prairie farms. In this 1982 photograph, a farmer near Tisdale, Saskatchewan, is using one for fall cultivation. (Wetherell Photograph)

local interest, he began manufacturing cultivators in 1970. His first units were 53-feet wide, solidly built, and constructed in five sections. Public reaction to the equipment was positive, and production was expanded that fall. By 1974, the plant was producing 300 cultivators per year plus trailers and rock pickers. By the early 1980s, Friggstad was also producing air seeders, a radically different type of seeding machine that used forced air to convey the seed into the ground. By this point, however, Friggstad Manufacturing was in financial difficulty. Like many implement companies, it found the early 1980s difficult because of a depressed farm economy, and in 1984, the company went into receivership. It was purchased by Flexi-Coil, a Saskatoon manufacturer of packers and harrow drawbars.

Friggstad's interest in large equipment was paramount and his company concentrated on machinery for large farms. In 1970, he developed and tested a prototype of an 80-foot-wide cultivator, then considered "the world's largest single-unit cultivator using a single hitch."[82] In 1974, the company developed a hydraulic tandem hitch to join two cultivators together to give a total width of 82 feet.[83] This emphasis on large equipment marked the sort of concerns that occupied large prairie grain farmers in the 1970s. Other companies also responded to these demands and the changes in equipment that they brought about. For example, in the late 1970s, Morris Rod Weeder Co. began production of chisel plows that were compatible with articulated four wheel drive tractors. The hitch was designed to allow the plow to turn tight corners.[84]

Cultivator design was not, however, only concerned with size. The issue of uniform depth remained significant and became critical with the development of air-seeding equipment which was used in conjunction with heavy duty cultivators. As will be discussed in a later chapter, ordinary cultivators with a rigid frame were not compatible with air seeding because the cultivator did not respect differing topography. The trick was to invent a cultivator that would "twist" with minor changes in the land. Olaf Friggstad's son Terry had researched and tested floating-frame cultivator technology, and this research was perfected at Flexi-Coil after its acquisition of Friggstad Manufacturing. The cultivator was built with separately floating sections. All frames of the cultivator flexed internally to maintain even depth with changes in the land. This was a highly sophisticated development and reinforced the use of huge equipment on large, minimum tillage farms for which the use of air seeders was logical.[85]

Different types of hitches and frames allowed heavy duty cultivators to "float." This provided better control and made them suitable for air seeding. (Flexi-Coil Photograph)

Trash Farming and Secondary Tillage Equipment

Another development that became significant with the increased popularity of trash-cover farming was the rod weeder. Although it had been patented in the United States in 1911, the rod weeder was still considered a new type of implement on the Prairies during the 1930s. Most rod weeders used a subsurface rotating rod that tore out tap-rooted weeds such as Canada thistle. Because the rod rotated opposite to the direction of travel, it flipped the weeds onto the surface. It also provided some soil packing and did not pulverize the soil. When used as an attachment with disc implements, it brought trash to the surface.

Given prairie conditions, interest in rod weeders was high. By 1919, a number of home-made versions had been developed since commercially produced ones were not yet available on the Prairies. Some of these, such as one developed by C.S. Noble in 1919, used a fixed rather than a rotating rod.[86] Others featured a system in which a wire or cable rod was strung between rotating drums. These wire or cable weeders were manufactured by the Riverside Iron Works, Calgary, from about the mid 1920s until the late 1940s as the Mills Wire Weeder. While it worked well on highly cultivated fields, and even on stony land, it clogged in fields with heavy trash cover. With increased adoption of trash-cover farming by the late 1940s, it faced a declining market.[87] Of all the rod weeders, revolving rod types proved most popular. Full line companies began marketing rod weeders by the 1920s, and there was considerable regional activity as well. The Tractor and Thresher Co., Saskatoon, manufactured the Jackson Rotary Rod Weeder in 1928, advertising that it performed a packing and mulching operation as well as weeding. Based on American designs, the Samson Rotary Rod Weeder Ltd., Calgary, also produced rod weeders which it sold across the Prairies from the early 1920s until 1962. Another prairie manufacturer of rod weeders was the Superior Rod Weeder Co. Ltd., Calgary. While it manufactured a sound product, the company failed because of a lack of financing and slow sales during its start-up years in the early 1930s.[88]

An early subsurface tillage implement used for secondary tillage was this Mill's Wire Weeder, manufactured in Calgary by Riverside Iron Works in 1927. (NA–3345–5, GAI)

The use of rod weeders was initially limited by the fact that the rod either broke or was thrown out of the soil when it hit a stone. There were various attempts to solve this problem, but none were successful until George Morris developed an automatic release mechanism. Morris was an implement dealer in Bangor, Saskatchewan, and farmers complained that the rod weeders he sold them operated poorly in stony soil. In response, Morris began work on an automatic spring-loaded trip device that permitted a section of the rod weeder to trip backwards when a stone or root was hit. The rod returned automatically to its position in the soil once the obstacle was passed. While competitors soon also featured trip devices, none had an automatic return. By 1929, Morris had patented the device. He initially approached the full line companies with his idea, but they turned him down—the trip device held no attraction for the American market.[89] Not to be stopped, by 1929 Morris and an assistant were producing eight-foot horse-drawn rod weeders featuring the trip device. In 1935, manufacturing was contracted to a firm in Portage la Prairie.

The Morris rod weeder was a success. As George Morris later recalled, "we were there with the right machine just at the right time when people were starting to wake up about how to farm the plains." [90] Ongoing redesign improved the equipment and met new needs. As amounts of trash left on fields increased, Morris adapted his equipment. In 1951, for example, clearances were increased to accommodate fields with heavier trash cover.[91] While some competitors used a partial wood frame, Morris rod weeders featured a sturdy all metal frame. A significant aspect of Morris's marketing was to provide solid guarantees for the machinery. He also provided purchasers with options that could be added in the future, announcing in 1939 that the "lower priced models can be converted to any model desired by adding the extra parts, that are sold as assemblies." [92] By 1942, the company offered three sizes of rod weeders (the largest was 12 feet); by 1960, its largest model had increased to 20 feet and a dual hitch was available to operate two 12-foot models as a single unit.[93] Within a decade, sizes reflected further increases in the scale of farming: the 1968 model was 36-feet wide with hydraulic wing lift to make transport easy.[94]

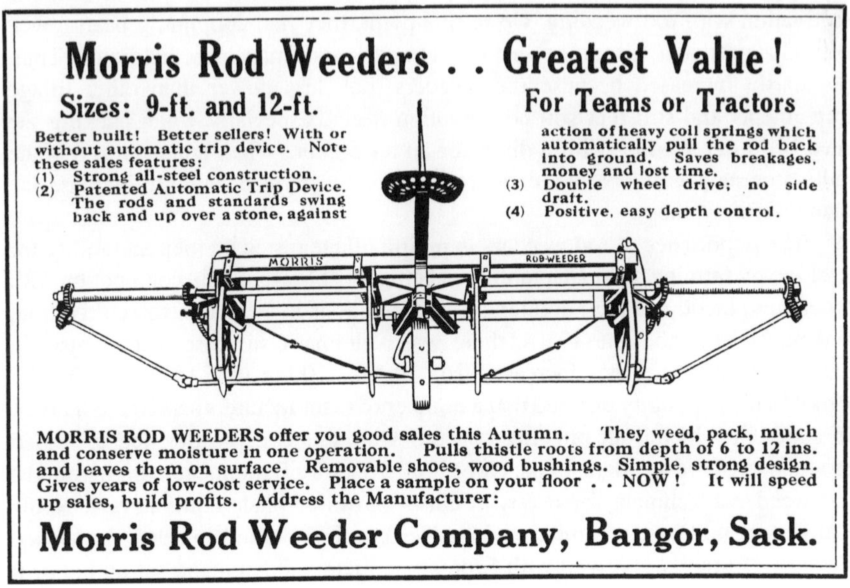

The Morris rod weeder featured a trip device that helped prevent breakage of the rod if it hit a stone or root. (*Canadian Farm Implements,* August 1936)

All manufacturing continued to be done at Portage la Prairie until the company opened a factory with about 40 employees in Yorkton in 1961. Two plants were also established in the United States to serve that market. By 1979, the company used 250 dealers in Canada, 175 in the U.S., and 30 in Australia, which was an emerging market. In the early 1960s, the company began a major expansion in its lines: adding a seed drill, a chisel plow, a light field cultivator, and then mounted harrows. George Morris's contributions to agriculture were recognized by a number of honours and awards, including membership in Saskatchewan's Agricultural Hall of Fame, the Order of Canada, and honourary degrees.[95]

A number of adaptations in rod weeders made them increasingly attractive to a broader range of farmers. One major design change was introduced by Francis Doepker of Doepker Industries in Annaheim, Saskatchewan. In 1963, he began manufacturing a rod weeder with a rod driven from the tractor power-take-off. This was a principle different than others on the market which had ground-driven rods. The AMA found that because the "rod was positively driven, the Doepker penetrated better than most rod weeders."[96]

Other design changes also allowed rod weeders to work in heavier stubble, making them effective as major implements for weed control on fallow land. As well, by the late 1940s the packing provided by rod weeders was found to be effective in controlling wireworms. In 1940, Morris Rod Weeder Co. began installing optional chisel points on its machines, and like other Morris developments, existing rod weeder models could be converted to the new system.[97] By about 1950, most rod weeders were equipped with cultivator shovels to combine cultivation with rod weeding. Up to this point, they had commonly been towed behind a cultivator, discer, or seed drills to pack and finish a seeded field.[98] Their popularity increased because rod weeders took less power than other tillage implements and still met soil conservation needs, especially when packing the seed-bed after seeding with a discer or an air seeder. When attached to a blade cultivator such as the Noble blade, rod weeders were also found to increase the blade's effectiveness.

The importance of rod weeders in prairie tillage rested on their suitability for trash-cover farming, their adaptability in being useable alone or in conjunction with other equipment, and their combination of functions in weed control, cultivation, and packing. In the late 1950s, there was a dramatic increase in their use: in Saskatchewan alone, they increased from about 1,100 in 1957 to about 4,000 by 1960.[99] This popularity ensured that a number of manufacturers served the market. In southern Alberta, an ongoing concern with soil conservation led to varied production of rod weeders. In the early 1960s, Victory Equipment manufactured a rod weeder attachment for use with chisel plows or blade cultivators.[100] Noble Cultivators produced a similar rod weeder attachment, and in 1965, Lawrence Edwards, formerly involved with Erdman in Victory Equipment, set up his own company in Lethbridge to manufacture rod weeders and rod weeder attachments. Edwards soon expanded and by 1972 was also producing chisel plows designed to take rod weeder attachments that were driven from the cultivator wheel. In this case, the trend to large equipment was not followed; these cultivators were suitable for middle-sized equipment and were available up to 21-feet wide.[101]

Through all of these changes in tillage, packing was one distinctive field function that continued to be as important in trash-cover farming in parts of the Prairies as it had been in black-soil summerfallowing. With trash-cover farming, packers were principally used to pack the soil after seeding, although in some districts they continued to be used after cultivation. By the early 1930s, packers continued generally to be either surface or subsurface types and used features common from the World War I period.[102] The drought of the 1930s increased the popularity of packers. On light soils, they were needed behind one-ways, especially when seeding, and farmers found subsurface types effective. In 1937, a number of firms were manufacturing them, and as was typical in so many other cases, many were made by iron foundries. In that year, Calgary Iron Works; Riverside Iron Works and the Samson Rotary Rod Weeder Co., both of Calgary; John East Iron Works, Saskatoon; The Prince Albert Foundry; the Lethbridge Iron Works; and Acme Manufacturing Co., Winnipeg, were manufacturing packers on a scheduled basis.[103]

There was some experimentation before World War II with different types of packers. C.S. Noble devised a subsurface packer with solid-cast wheels for use behind seed drills. But more important developments took place after the war. By 1948, packing was still found to be necessary for moisture retention in conjunction with seeding, and for control of wireworms. But surface packers, which contributed to soil drifting, were losing popularity in favour of subsurface and other types of packers that would work better with trash cover. Among the most significant developments in this respect was Emerson and Ken Summach's coiled land packer. In the late 1940s, Emerson Summach, a farmer near Asquith, Saskatchewan, developed a packer that consisted of a single steel coil about 18 inches in diameter. The coil had no spokes and was simply a continuous piece of coiled steel. He reasoned that such a design would not plug like the packers then in use. Farming was paying poorly, and Summach saw his idea as a route to a better income.[104] He approached several full line companies with the concept, but they rejected it since there was no existing machinery to turn out the coils. In response, Summach designed a machine to roll steel into a continuous coil.[105] This combination of product development and tooling for manufacturing was crucial. Summach moved to Saskatoon and set up a small plant and a company which he called Flexi-Coil. Packers were marketed through advertisements in newspapers and magazines. By 1953, a few dealers handled the packer, and given interest in the United States, manufacturing and distribution there were contracted to an American firm.[106]

The coil packer was a simple but ingenious design with significant advantages over existing types. It followed the contours of the land, it worked well in stony land, and it did not pulverize the soil. Moreover, as was noted in one report in 1955, because it had no spokes, it did not clog up in ordinary soil, and even if a large lump of mud worked its way into the coil, it "simply 'augers' its way out to the end." In effect, it could be used behind any sort of cultivation equipment. While it levelled the soil, it produced diagonal ridges on the soil (instead of straight ones), which further contributed to soil conservation.[107]

By 1957, Flexi-Coil had produced 4,000 packers.[108] Emerson Summach died in 1964 and his place in the company was taken over by his son Terry. The company grew steadily and in 1966 began manufacturing a harrow packer drawbar, its second line of equipment. Flexi-Coil continued to expand, but by the early 1980s, its markets were slipping under pressure of hard economic times. While it remained in better shape than companies producing more expensive equipment like tractors, Flex-Coil nonetheless felt the effect of rising interest rates and the general recession. At this point, Flexi-Coil purchased Friggstad Manufacturing, which was then in receivership, gaining Friggstad's technology in air seeders and providing the basis for significant expansion for the balance of the 1980s. The company then began to make not only packers but heavy duty cultivators, harrows, spraying equipment, and, of course, air seeders.[109]

The Summach brothers of Asquith, Saskatchewan, designed and manufactured a unique packer that promoted soil conservation. Here it is shown being manufactured at their Saskatoon plant in 1961. (Flexi-Coil Photograph)

Like packing, harrowing evolved with the changing nature of prairie tillage. In the early 1930s, the common types continued to be spike tooth (also called peg tooth), spring tooth, disc, and chain harrows. All were used to control small weeds, to level and pack the soil during fall tillage, and, in some areas, to kill grasshopper eggs and wheat stem sawfly grubs. Concerns about disc harrows and spike tooth harrows were, however, beginning to be expressed. Although disc harrows by then saw decreased use, Saskatchewan farmers were warned in 1933 to use them and spike tooth harrows with "care" to prevent conditions favourable to soil drifting. Chain harrows were even worse in these terms, but since their primary use by then was for shaking out couch grass brought to the surface by a cultivator or a spring tooth harrow, they were too specialized to cause general concern. Spring tooth harrows, on the other hand, were increasing in popularity. Used mainly for shallow cultivation and for killing creeping rooted weeds like couch grass, they were favoured because they left a lumpy surface.[110]

By the 1940s, many of the same concerns about the use of harrowing equipment remained.[111] However, probably more than in any other type of field equipment, there was an increasing variety of harrows to choose from. New designs and new features were continually introduced. Among the unique new designs for harrows was a self-cleaning "harrower" introduced by CCIL in 1946. It was the first equipment manufactured by the company and utilized the patented design of Elmer Carlson of Wetaskiwin who allowed CCIL to use it free of charge. The harrower had four concentric rings placed within one another. It was towed by a drawbar mounted on wheels. While rotary-type harrows were not new, earlier designs had either tended to clog or required too much power to operate. Carlson's design avoided these problems and was advertised as ideal for levelling and shallow cultivation on trash-covered fields. It also worked some straw into the soil. While significant in technological terms, the harrower was not the success CCIL anticipated. The machine did not stand up well and so lost much of its popularity.[112] CCIL did, however, meet with greater success when it introduced a folding harrow drawbar. It was designed by Helmer Hanson of Lajord, Saskatchewan, a farmer who had invented the swather in the late 1920s. The drawbar was innovative because it permitted easy transport. It replaced rigid-frame drawbars, and the concept was widely copied by other manufacturers. Folding drawbars with hydraulic lifts resulted in further convenience and mobility. Francis Doepker developed a hydraulic drawbar in 1956, a principle that came into general use in the next decade.[113]

Throughout the 1950s, spring tooth harrows continued to be popular and they saw even greater use by the early 1960s. Using long tines, they provided shallow cultivation in trash-covered fields and in preparation of the seed-bed. While they had most often been used alone, by the early 1960s, they were commonly attached to a heavy duty cultivator. One of the first prairie firms to do so, Crown Industries of Regina began manufacturing attached spring tooth harrows in about 1960.[114] By this time, there was an ever increasing array of harrowing equipment, much of it produced by metal-working companies. For example, by the late 1970s, Inland Steel and Forgings Ltd., Winnipeg, was one of the largest harrow manufacturers in Canada.[115]

While changes in tillage practices were forced by soil degradation, they were also the result of a broad shift taking place in prairie farming. Aside from soil conservation, these changes offered solid economic advantages that help to explain their attraction. Their adoption throughout the Prairies, even in areas (such as the park-belt) where soil drifting was not a problem, is evidence of this factor. Evan Hardy noted this in 1927 when pointing out the lower operating costs of cultivators and disc implements over plows.[116] In 1932, it was also observed that "just a few years ago" the plow and the harrow were seen as the "most important," and often the only tillage equipment necessary on prairie farms. Now, the drag harrow was fast disappearing and moldboard plows were being replaced with the duckfoot cultivator and the one-way disc. As well, rod weeders were favoured in many districts. These changes represented "a changed conception of what constitutes good tillage." Equally important, they were cost effective because they combined operations, and they coincided with "the necessity of speeding up tillage operation to keep up with faster methods of seeding and harvesting."[117]

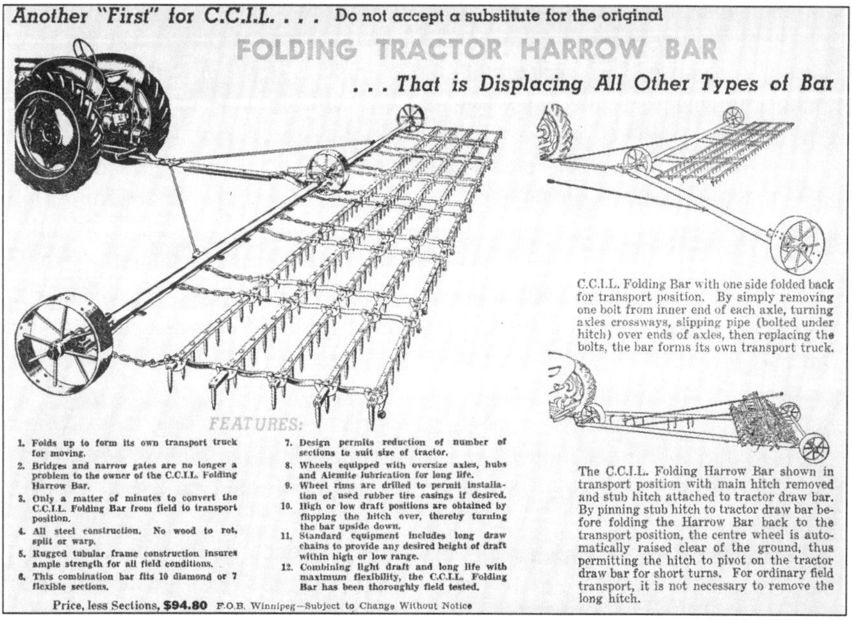

Increased size of farms created a need for large equipment that was easy to transport. This folding harrow drawbar, manufactured by CCIL, was based on the designs of a prairie farmer. (*CCIL Farm Machinery Catalogue* 1950)

The economic rationale for changes in tillage practices was therefore strong, and it fit neatly with new methods of soil conservation. The climatic problems of the 1930s combined with the increased size of farms and a desire to save costs by lessening field operations had by the mid 1940s changed tillage practices on the Prairies. By 1955, field tillage tended to emphasize trash cover and a "cloddy" surface which could be produced by heavy duty cultivators. Even so, concerns about soil conservation could not be forgotten. Some farmers continued to use black summerfallow, even in drier areas. And, as speed of operations increased, especially with large heavy duty cultivators, residue conservation decreased. Nevertheless, the new approaches were widespread. In very dry areas, subsurface tillage equipment like the Noble blade, strip cropping, and other techniques were also used to prevent erosion and conserve moisture. All of these changes created new approaches; by the 1970s, most of the tillage equipment that had been in use in the early 1920s had been replaced with different equipment, much of it manufactured by prairie short line firms.

Seeding, Fertilizing and Applying Chemicals

The benefits of combining field operations to save money and time created an interplay between seeding and tillage. Equally important in creating this relationship was the need for soil conservation, and the development of trash cover farming influenced the technology of seeding. While good crops depended on good tillage, correct seeding of grain was also of central importance, as were various other techniques for stimulating growth and controlling damage from insects and weeds. Accordingly, cleaning seed, applying chemicals to control weeds and insects, and using fertilizers to stimulate growth were integral elements in crop maintenance.

Seeding machines were used by the Chinese by at least 2800 B.C. and by the Babylonians about 2000 B.C. In Europe, invention of seeders began in sixteenth century Italy, but only in the eighteenth century—with agricultural experimentation in France and England (notably by Jethro Tull)—did it move beyond the experimental stage. Most of these seeding machines were drills, that is, machines for sowing crops in rows. In contrast, a dominant trend on the twentieth century prairies favoured seeders, that is, machines that broadcast the seed beneath the soil. But both seeders and drills shared the similar goals of placing seed in a way that would ensure rapid germination and increased yields. On the Prairies, an especially important part of this goal was to seed at an even and sufficient depth to reach moisture for germination. Within this broad objective, there was some debate about methods, especially after World War II. Some claimed that precisely calculated spacing of seed and depth could increase yields, while others argued that the lower yields produced by more inaccurate systems were offset by their lower cost.[1]

Seeding with Drills

Seed drills had come into common use in the American west by the 1880s, and almost all seeding in Manitoba was done with seed drills by the late 1880s. The drills used on the frontier would have featured some type of forced feed, with the commonest furrow openers being shoe or hoe types. They required a carefully prepared seed-bed since they did not cultivate for weed control, and they plugged easily with heavy trash. Stubble had to be plowed under at least six inches and the soil had to be packed to create a firm seed-bed. By about 1890, press wheels were attached behind each opener (making it into a press drill) to press the soil down over

Early seed drills required that the field be finely cultivated. This farmer is seeding at Coaldale, Alberta, in 1918. (Harry Pollard Collection, P622, PAA)

each row of seed. This worked well in areas where early and uniform germination did not occur easily, and by 1920, press drills were in general use in the prairie dry belt.[2]

Throughout the interwar years, shoe drills and hoe drills remained in use on the Prairies. Shoe drills worked best in light soils that were "reasonably free from trash." Hoe drills were more adaptable machines and could also be used on hard or rough soils "where it is desired to have the soil in rough cloddy condition."[3] While shoe drills with press attachments continued to be used, they were gradually replaced with hoe press drills which produced good yields on light soils, especially in dry years. Hoe press drills achieved good seed placement because they provided good penetration in hard ground and worked in fields with "considerable" dry trash. This ongoing popularity was ensured through design changes. C.S. Noble recalled that in the early 1920s, when soil drifting was common, hoe-type drills had been best for seeding because they produced even germination and because the furrows created by the press attachment protected the plants. However, the growing acceptance of the need for trash cover by World War II made this type of hoe drill obsolete since it lacked clearance to seed through heavy trash.[4]

Significant work took place in southern Alberta in developing hoe drills that worked in trash-covered fields. In 1951, Noble Cultivators Ltd. introduced a high clearance hoe press drill as part of its ongoing efforts with soil conservation equipment. The hoe cut a furrow in which the seed was deposited and the rows were packed with a press attachment with solid wheels "to avoid the dust nuisance caused by open spoke packer wheels." The stubble was left standing, providing protection to the seedlings and conserving moisture.[5] The other important southern Alberta development in this type of equipment involved the Felgar Hutterite colony. Shortly after Noble brought out its high clearance drill, the colony was given a demonstration of its operation. They liked its principles, but wanted the rows more closely spaced to create a thicker stubble to hold up the swath. Along with the local John Deere dealer and other company representatives, modifications of a John Deere press drill were undertaken at the colony's machine shop. The redesign resulted in the successful John Deere Model LZ mulch-hoe-drill.[6]

The leadership shown in southern Alberta in the development of high clearance hoe drills did not preclude developments elsewhere on the Prairies. In 1961, Morris Rod Weeder Co. developed a prototype for its 13-foot-wide Seed Rite hoe press drill for seeding into stubble. Like other Morris equipment, it featured an automatic trip device on the shanks for easier use on stony land. While it was not a novel principle, given Morris's work with rod weeders, the Seed Rite also featured a rod weeder attachment to combine tillage and seeding.[7] In this system, furrows were opened by a hoe-type opener. Immediately behind the openers was a power-driven rotating rod. It passed about one inch above the seed to compact soil around the seed and destroy tap-rooted weeds as well as those growing in the soil above the seed. It left a loosely textured surface on the field, although the height of the rod had to be adjusted correctly to be certain that enough soil, in which trash could be anchored, passed over the top of the rod. This system was therefore difficult to use for shallow seeding, and if the soil was hard, the rod weeder could loosen lumps of soil which later had to be packed or harrowed.[8]

In 1951, C.S. Noble brought out this high clearance press drill to seed trash-covered fields. (Noble trade literature, ca. 1959, GAI)

Throughout the 1970s and 1980s, there was still a substantial demand for high clearance hoe drills suitable for minimum-tillage farming in arid conditions. Morris Rod Weeder Co. increased its role as a manufacturer of this seeding equipment when it purchased Terra Implements of Minnedosa, Manitoba, which manufactured press drills and tandem discs under the brand name Terra.[9] As well, New Noble Services Ltd. produced the Noble press seed drill, and Edwards Grain Guard, Lethbridge (established in 1965 as Edwards Rod Weeder Co.), also produced high clearance drills in the 1980s. Their model came with or without a fertilizer attachment and had optional rubber packer wheels. Similarly, Cereal Implements (the division of Vicon manufacturing the CCIL lines) put out a drill that was advertised as having sufficient depth and clearance "to punch through some of the heaviest trash and toughest soil." Indeed, these drills could "be used as a once over seeding operation" and weeds could be sprayed after seeding.[10]

Despite their continuing popularity, hoe press drills were increasingly challenged after World War I by disc drills. Although these drills had been developed in the late nineteenth century, they did not become common on the Prairies until the interwar years. They did not provide a complete tillage function and therefore required the preparation of the seed-bed by cultivation. Discs were used to open a furrow, and the drills placed seed at a uniform depth and covered it without loosening the soil too much. This feature, plus their suitability for light soils, made them a popular choice in areas subject to soil drifting and aridity.

Disc drills came in either single or double disc types. Single disc drills featured a series of individually suspended curved vertical discs which were pulled at a slight angle to open a furrow. Essentially, they were disc harrows with a seed box attached, and they were often used on land that was difficult to seed because of roots or heavy stubble. But, reflecting their true nature as a disc harrow, they tended to pulverize the soil, and by the early 1960s, they had become uncommon. In contrast, double disc drills did not overly pulverize the soil and so remained popular. They had two discs placed vertically so that they touched at the front to cut a furrow. By the interwar years, the common practice of dragging chains behind the drill to be sure that the seed was covered was discouraged because it pulverized the soil and contributed to drifting. Instead, press wheels were used to pack the soil over each row. The ridges left by the press attachment also helped prevent soil drifting.[11]

Disc drills were commonly used throughout the interwar years, and by 1948, they were still "the standard" drill in Saskatchewan. Double disc drills were most common in Manitoba and Saskatchewan, while single disc types were most common in parts of Alberta. One major change in these drills was in size. In 1932, it was observed that "a few years ago" a 20-run drill had been standard but now it was considered a minimum. By 1950, most drills could seed about 40 acres per day.[12] Most of these disc drills were manufactured by full line companies, but some were produced on the Prairies, including one by New Noble in the late 1980s.

Seeders

The continued popularity of drills did not prevent the development of alternate systems of seeding. Seager Wheeler had written in 1919 that "there is great need for an improvement in seed drills." Indeed, he favoured seeders over drills because he believed that drills crowded the seeds in the row. "We want the seed broadcast under the ground in three or four inch rows instead of a narrow row." Otherwise, the seed was "crowded and the weaker plants suffer" and weeds became established between the rows.[13] Another disadvantage of the drill was that while it provided some tillage, separate cultivation was still necessary to prepare the seed-bed. Before World War II, there were a variety of efforts to combine drilling and tillage into a single operation, but none were particularly successful. They had included towing small seed drills behind a moldboard plow, a one-way, or a cultivator. A packer was "generally necessary between the press drill and the plow or one-way to control the depth of seeding and produce a firm seedbed." While this served to combine plowing, drilling, and packing into one operation, it still retained the advantages of the drill, especially in dry conditions.[14] The logical step, of course, was to combine all of these functions within a single machine.

This Alberta farmer combined plowing, packing, and seeding with a press drill into a single operation during the 1930s. (Archives Collection, A7763, PAA)

Among the earliest examples of such a machine was the Mitchell sub-seeder, invented by J.A. Mitchell of Lawson, Saskatchewan, in the early 1920s. Mitchell's seeder used a duckfoot cultivator blade to open the seed-bed and the seed was broadcast from an attached seed box. This system broadcast the seed under the ground, which Mitchell argued gave each plant a "better chance to develop." Equally attractive, it would work on summerfallow stubble with "no prepatory cultivation such as is usually necessary for present makes of drills." Mitchell wrote that his seeder

> will cultivate by seeding, and will spread the seed in a wide row at almost any depth desired. It is kept in the ground by suction or draw caused by its sharp edges...and will, therefore, run at a more uniform depth in changeable land. The dry earth from the surface has little or no chance to get next to the seed. You can seed more grain per acre and not cramp the plants. It is not so good in loose plowing that has weeds or stubble plowed in, but is intended for land that needs cultivating.[15]

By 1929, the Mitchell sub-seeder was being manufactured and sold by the Tudhope-Anderson Co. Ltd., Orillia, Ontario, and was being used by a number of prairie farmers, especially in light soil areas in Saskatchewan. While it was an important innovation, it was found that in soils where the cultivator did not "scour and clean satisfactorily," it was difficult "to get uniform covering, tending to leave the soil in such a condition that it will dry out badly." The result was incomplete and irregular germination.[16] In a similar development, in 1930, Winnipeg's

Kirchner and Co. was manufacturing three models of "a combination of plow, seeder and cultivator." A seed box was mounted above either four- or six-plow bottoms. Seed tubes carried the seed into the furrows, "scattering the grain over the whole width of the furrows directly in rear of the shears." The seeder plow also featured a device for manually regulating depth to ensure uniform seed depth.[17]

Neither the Kirchner seeder plow nor the Mitchell sub-seeder were wholly satisfactory for the Prairies. Tests in 1931 at the Swift Current experimental station found that the Kirchner machine "did not produce a very good yield" because the moldboards pulverized the soil and left it too open, resulting in high loss of moisture. Similar problems were found with the Mitchell sub-seeder, and the experimental station concluded that better yields could be had in dry conditions by using most types of conventional drills.[18] Use of such seeders nevertheless seems to have endured to some extent, although the problems of excessive seed-bed drying and uneven depth of seed placement remained unresolved.[19]

Other efforts in this direction included C.S. Noble's experiments mounting a seed box on a blade cultivator. Of these attempts to combine tillage with seeding, the most popular method was to mount a seed box on a one-way and, later, on a discer. This allowed the farmer to seed and cultivate in a single operation, including seeding stubble land.

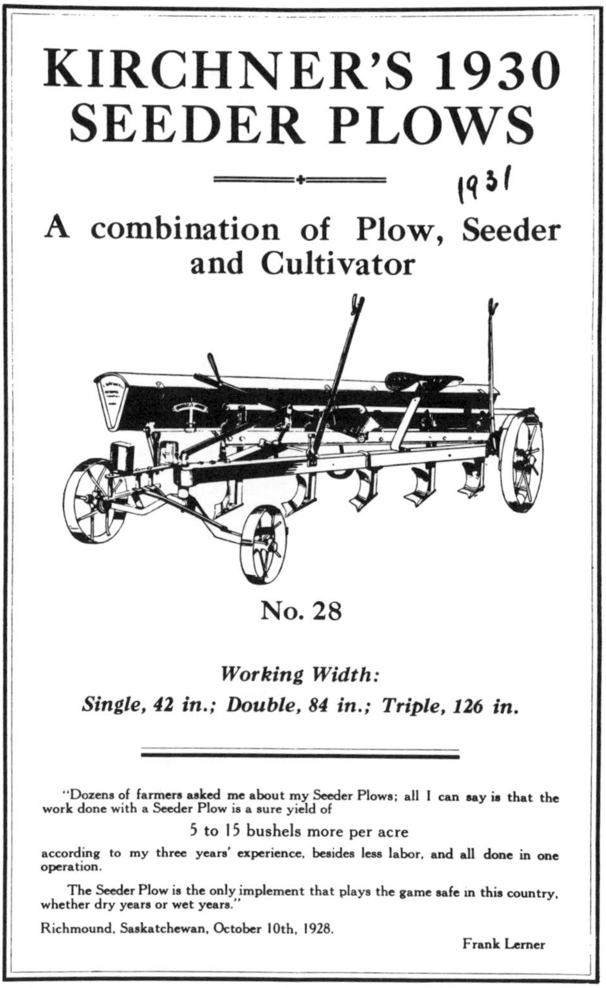

An early effort to find an alternative to seed drills and accommodate changing field practices was the Kirchner seeder plow in 1930. (Ag. 12. II. B. 39, SABS)

Seeding stubble land was a logical development because of the growing concern about soil conservation. While there were initial problems with irregular emergence of seedlings, by 1931, improved designs of one-ways with seeder boxes resulted in more even growth. They were found to produce higher yields on stubble than other methods, including those requiring pre-seeding tillage, and on summerfallow, they produced yields equivalent to those obtained with drills. Since one-ways and discers applied the same principles, these observations applied equally to discers, and by 1970, they were judged to be "probably the most widely used machine on the Prairies."[20]

As the soil was cut by the discs on a one-way or discer, the seed was broadcast onto the bottom of the furrow and covered by the soil being inverted by the disc blades. On light soils and stubble, this often left the soil loose, and packing was needed both to achieve rapid and uniform germination and to prevent erosion. However, the one-way or discer with a seed box worked best on clay soils, and many farmers found that conventional drills remained the better choice for light soils because seedlings emerged more uniformly. On uneven fields, they produced variable seeding depth, so that the field had to be harrowed before seeding. Still, they had significant advantages. They successfully combined tillage and seeding and could be used in wetter conditions than drills. They also combined equipment use since the discer could be used for summerfallowing.[21] The approach of using a one-way with a seed box was apparently first devised by farmers. Later, the idea was largely picked up by full line implement manufacturers,[22] although some prairie short line companies did manufacture them. CCIL produced discers with seed boxes during the 1950s and 1960s, and this production was continued by Vicon. By the late 1980s, Vicon produced units in 12- and 15-foot widths and multi-unit widths of up to 60 feet.[23]

Air Seeders

After World War II, significant change in seeding practices began with the use of forced air (that is, pneumatic systems) to carry the seed into the soil. The air seeder was perhaps the most important technological change in prairie agricultural equipment after World War II, and much of its development took place on the Prairies. One of the earliest air seeders was produced by Victor Erdman of Victory Equipment Ltd., Lethbridge, and it was tested at the Swift Current experimental station in 1952. It consisted of a seed box mounted on a Victory blade cultivator. A gasoline-powered blower forced the seeds along tubes and beneath the rear of the blades and deposited them in wide rows, or bands. Mechanically, this system worked well on level land, but it planted too deeply for even and early germination.[24]

Early efforts like Erdman's did not immediately change seeding technology. During the 1950s and 1960s, there were few commercial efforts to develop air seeders on the Prairies, although interest in them was high. By the 1970s, however, a number of air seeders had reached the marketing stage. For example, Pride Industries of Virden, Manitoba, brought out its Super Seed Air Seeder in 1977. The

Among the first air seeders designed on the Prairies was this one by Victor Erdman of Lethbridge. It was tested at the Lethbridge Research Station in 1952. (Agriculture Canada Research Station Photograph, Lethbridge, Alberta)

machine, which had been field tested for five years, had been developed by two Saskatchewan farmers, Preston Davy and Duncan Elliot. They, along with Richard Evanson of Virden, formed Pride Industries to market the machine. They argued that their air seeder provided more precise placement of seed and therefore resulted in better germination. Because the seed-bed was prepared at the same time it was seeded, it was also well suited for the minimum tillage methods in use in many parts of the Prairies. The machine was built for Pride Industries by Premium Developments, the same firm that manufactured some equipment for the Morris Rod Weeder Co.,[25] and it had a number of features that came to define air seeders in later years.

Unlike Victor Erdman's system of the early 1950s, the air seeders of the late 1970s had a separate seed tank that was towed along with the cultivator. While the tank on these machines had separate compartments for seed and fertilizer, only liquid, not granulated, fertilizer could be used. The seed and fertilizer were forced from this tank through long tubes to the cultivator shovels and broadcast in the ground in a four- or five-inch band. The separation of the seed box from the cultivator resulted in better stability and control of the cultivator, making it easy to transport. This air seeder also permitted seeding into more trash than did conventional equipment. And, since the soil was opened with a cultivator, it also helped control weeds, and tillage in preparation for seeding could often be dispensed with. However, the seed-bed had to be firmed with a packer.[26]

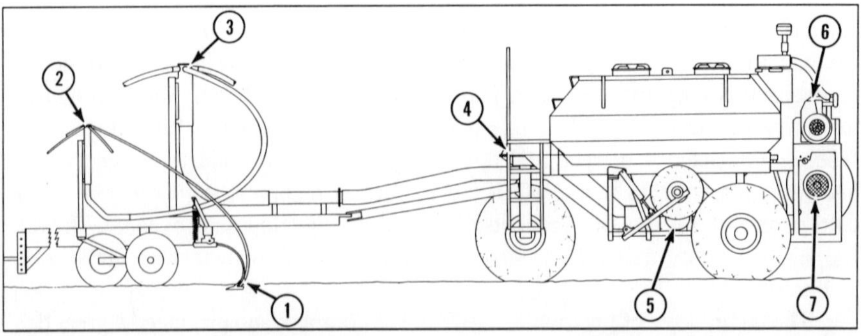

This schematic drawing of an air seeder produced by F.P. Bourgault Industries, St. Brieux, Saskatchewan, illustrates the operation of this equipment. Bourgault Model 138D Air Seeder: (1) Seed Boot, (2) Secondary Header, (3) Primary Header, (4) Tank Access Ladder, (5) Metering System, (6) Engine, (7) Fan. (PAMI/Alberta Agriculture Photograph)

By 1980, further developments had taken place in air seeders. They could now use granular fertilizers and could seed a wide area (up to 90 seed runs) at up to eight miles per hour. In 1981, the Lethbridge station of PAMI tested several new air seeders, including one by Friggstad Manufacturing and two by Prasco Super Seeder Ltd., Winnipeg. While PAMI found that the machines functioned well, some general problems remained. Most importantly, uniform depth of seeding was difficult to obtain and adjustments to change seeding rates were awkward.[27] These problems were overcome in the 1980s, making this a decade of significant development in air-seeding technology.

By the 1980s, there were about a dozen firms in Canada manufacturing air seeders. Among the most successful of the first of these manufacturers was Bourgault Industries of St. Brieux, Saskatchewan. As a manufacturer of cultivators, Bourgault's expansion into air seeders was a logical step. Inspired by the use of air seeders in dry-land Australian farming, Francois Bourgault began work on air-seeding technology and by 1981 was manufacturing an air seeder with a ground-driven metering system.[28] By the late 1980s, many of the major prairie equipment manufacturers were also producing air seeders. The technology they used was a mix of prairie and international invention. Some companies like Leon's Manufacturing and Ezee-On Manufacturing purchased designs for air seeders, as did Friggstad Manufacturing, which began with some German-designed components. Flexi-Coil's 1984 purchase of Friggstad Manufacturing brought tillage equipment to Flexi-Coil's product line, and marked the beginning of major expansion for the firm. While Flexi-Coil had marketed an Australian air seeder starting in 1980, it had no tillage to go with it. Its acquisition of Friggstad's technology, however, allowed it to proceed with its interest in air seeding.

In all seeding systems, consistent depth of seed placement was crucial to achieve relatively even ripening of the crop. Precise placement of seed was desirable to save seed, an especially pressing concern with expensive hybrid seed. These needs gave rise to the process of "banding," in which the cultivator cut the soil and the seed and fertilizer were deposited from the seeder tank through separate spouts into the soil in distinct bands. Each of these bands was layered into the soil in precise relationship to the other. The positioning of these bands could be changed to suit particular needs. While this precision was one potential advantage of air seeding, consistency of seed and fertilizer placement still had to be resolved. It became apparent that the solution lay in further technical development of the cultivator and not the air seeder itself. Thus, the development of large, flexible frame heavy duty cultivators played an integral part in furthering air seeding technology.[29]

The technology of pneumatic seeding was complex and produced a great deal of activity among some prairie manufacturers. It also created a number of spin-off opportunities for other regional manufacturing, such as the production of air blowers and monitors for air seeders. This was significant not only in illustrating the technical complexity of air seeding, but it also demonstrated the interplay that had developed among prairie manufacturers. In some cases this common interest resulted in mergers. In 1989, for example, Field Technologies Ltd., a manufacturer of monitor systems, was acquired by and became a subsidiary of Flexi-Coil.

Air seeding combined tillage and seeding to an unprecedented degree. It also allowed large scale operations as shown in this photograph from the late 1980s. (Flexi-Coil Photograph)

Cleaning and Treating Seed

While tillage practices were of central importance for weed and insect control, other techniques were also widely used. The most important of these included the burning of stubble and straw on fields heavily infested with weeds, sowing clean seed, and the use of herbicides and pesticides. The problem of weed control was a general and long-standing one. The territorial government passed a *Noxious Weed Ordinance* in 1898, and all provinces enacted similar legislation to help control the spread of weeds. In 1911, the federal government required that all grass, clover, and alfalfa sold for seed be graded, and a similar provision was extended to all grains in 1923. These regulations were difficult to enforce though and met some public resistance. By the 1930s, farmers were being encouraged to clean seed grain through government-sponsored seed cleaning programs and locally organized seed cleaning associations. In these cases, cleaning equipment was provided or subsidized by government.[30] By 1944, Alberta operated four such plants and Saskatchewan had a large plant at Moose Jaw.

Cleaning seed was of obvious importance for field production and reflected two different concerns: the removal of weed seeds and the control of diseases. By 1930, there were a number of grain cleaning attachments that fit onto threshing machines and combines, but they found only limited use on the Prairies. Farmers were unwilling to purchase them because they were not interchangeable among different brands of threshers or combines.[31] Instead, farmers marketed dirty grain and paid the dockage and freight costs. Few farmers had the time, the machinery, or the storage space to clean grain before marketing it. Instead, they usually cleaned only seed grain, most often with a fanning mill.[32]

There were at least four types of cleaning equipment commonly used and manufactured on the Prairies. These included fanning mills, oat kickers, carpet cleaners, and disc and cylinder cleaners and graders. Fanning mills were the earliest and commonest equipment used for seed cleaning and grading. Available in various sizes, all operated with sieves or rollers and a blast of air. When the sieves were shaken, the grain was separated by size. Simultaneously, air blown across the grain separated out by weight any light material and diseased kernels. Sieves came in various sizes for different needs, and the fanning mills were powered by hand or by gasoline engines. The grain needed to be put through the fanning mill twice to be properly cleaned.

The other three cleaners were less popular. Oat kickers, like fanning mills, also used sieves to separate grain by length, but they were fixed at a sharper angle than in a fanning mill. The grains of wheat passed through "the throat of the sieve, while the oats, being too long . . . are kicked backwards in the direction of the throw of the sieve" and out of the machine. With carpet cleaners, grain was passed over a moving fabric belt which caught those seeds with beards (such as wild oats and barley). These seeds were then brushed off. Disc and cylinder cleaners and graders all operated on similar principles. They were usually used in large commercial cleaning mills, although small models suitable for the farm were available after about 1925. Disc machines consisted of a series of rotating discs with small pockets of different sizes. As the grain moved through the machine, kernels of varying length were caught in the different pockets and were discharged from the machine. Cylinder cleaners operated in the same way, except that the pockets were fixed along a cylinder. Later, specific gravity separators were developed and these separated the seeds by differences in density.[33]

Fanning mills were one of the earliest machines for cleaning grain, and remained in use on prairie farms until after World War II, as this 1942 Alberta photograph shows. (Archives Collection, A6483, PAA)

In addition to ridding grain of weed seeds, this equipment was used to try and control smut. This disease was of major concern on the Prairies and in 1927 alone was estimated to have cost Canadian farmers $12.5 million. Although fanning mills were used for this purpose, they were of limited practical use since smut is a fungal disease. A better form of control was to treat grain with mercury dusts or formaldehyde. Formaldehyde, the oldest form of treatment, was diluted with water and the formaldehyde solution was sprinkled on the grain as it was being shovelled around. This task could also be accomplished by a machine called a pickler. There were many types of picklers available, ranging from relatively simple dunking mechanisms to more complicated systems in which the grain was coated as it was being fed through the machine. One feeding mechanism consisted of an auger, a system invented around 1900 by Robert Davidson, a Manitoba farmer from the Carberry district.[34] Since the grain became wet during these treatments—and so spoiled quickly—the treatments were usually carried out the evening before seeding.[35]

An alternative to these wet treatments was treating grain with copper carbonate or mercury dust. But because both of these chemicals, especially mercury dust, were toxic to mammals, they were used in moderation. The dust also clogged up the gears of seed drills. These factors were offset by the fact that dust treatments, unlike wet ones, could be done well in advance of seeding, which meant less work in the spring. While there were home-made applicators for dusting, commercially made machines were most effective and were generally used for applying dry chemicals.[36]

Before 1945, grain cleaning and treating equipment was among the most common product manufactured by prairie agricultural implement firms. This was perhaps encouraged by the special tariff protection they received, which made imported models more expensive. However, they were an ideal short line product since they required only small manufacturing facilities and relied on a specialized but large market. There was especially dramatic growth in the number of manufacturers of this equipment during the World War I period. This equipment was a common sideline for machine, iron, and tin shops, and for general manufacturers such as the Brandon Implement and Manufacturing Co. Ltd. In addition to machine repairs, it also manufactured a wide range of products including picklers, wagon boxes, packers, furnaces, and saws. As well, the Brett Manufacturing Co., Winnipeg, added picklers to its production line in about 1914, which until then had been dominated by incubators, brooders, and other farmyard equipment.

For other more specialized firms, picklers and cleaners were the major product. Beeman and Co., an American company that manufactured the "Jumbo" cleaner, set up a Canadian branch plant in Winnipeg in 1902. It advertised itself as the only American manufacturer of such equipment with a western Canadian plant "for the convenience of the Canadian trade." The Saska Manufacturing Co., Saskatoon, also marketed a grain cleaner from 1914 to 1916.[37] While heavily concentrated in cities, there was enough activity in this field to justify the existence of manufacturers in smaller prairie centres. The Farmers' Machine Co. Ltd., Watrous, Saskatchewan, manufactured a fanning mill at least from 1913 to 1916. Similarly, the Currie Manufacturing Company, Lauder, Manitoba, manufactured a pickler in 1915.

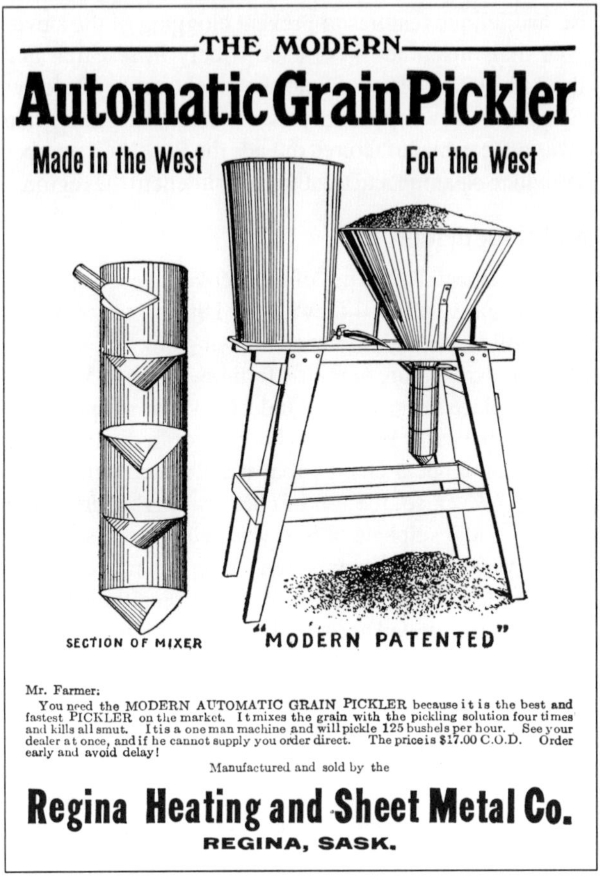

Grain picklers were an early short line product on the prairies. They were used to treat grain to control smut. (*Nor'-West Farmer,* February 20, 1915)

Although not many of these manufacturing efforts lasted for more than a few years, there were exceptions and some of these formed the basis for future expansion. In 1949, Mike Babysh and Eugene Demkiw of Vegreville began manufacturing a blanket fanning mill under license from John Phillips, a local inventor and farmer. The mill was unique in that it could separate wild from domestic oats. This manufacturing effort expanded into other lines, becoming Ezee-On Manufacturing in 1966. Other examples include the Emerson Manufacturing Co., Winnipeg, which began manufacturing its various brand name grain cleaners around 1918. An American branch plant, the company was reorganized as a wholly Canadian-owned company in 1960 and introduced a commercial cleaner to the market.

Carter Day Industries (Canada) Ltd., Winnipeg, was another manufacturer of a well known brand name grain cleaner, as was Forever Industries, Winnipeg. Specializing in grain cleaners, Forever Industries began operation in 1927. While complaints about poor quality and design of cleaning equipment (especially fanning mills) were common, Forever's machines were sturdy and made completely of metal.[38] They also had unique features to prevent clogging of the sieves and feeding mechanism, and their machines were used widely in research and government agencies. By the mid 1950s, the use of field chemicals resulted in fewer weeds in crops, and cleaning equipment was primarily used to clean and grade grain for seed.[39] While most of it was by then manufactured outside the Prairies, a number of companies continued to produce cleaning and treating equipment in the region.[40]

Chemical Herbicides and Insecticides

While changing tillage practices, especially the use of trash cover, were important for soil preservation, they had a negative effect—they allowed the growth of weeds that had previously been controlled through summerfallowing. In this context, chemical herbicides became a standard feature of prairie farming by the 1950s. This paralleled overall changes taking place in agriculture; indeed, the use of chemicals in North American agriculture after World War II has been described as a further element in the revolutionary technological change that transformed agriculture after the mid nineteenth century.[41] Even so, the use of chemical herbicides had a long tradition on the Prairies. By 1908, sulphate of iron was being used as a spray to control some field weeds, and various experiments were being undertaken to test its effectiveness. By the 1930s, soil sterilants were also in wide use, although their effectiveness for field applications was limited because they indiscriminately killed all plants. The use of chemical herbicides changed radically in 1945 when the first tests were carried out on a new herbicide, 2,4–Dichlorophenoxy-acid, or 2,4–D. By 1947, a half million acres on the Prairies were treated with 2,4–D, and by 1960, seven million acres were sprayed with herbicides, mainly 2,4–D.[42]

Developments in insect control through chemical means generally paralleled those in the area of weed control. From the late nineteenth century, insecticides were made from nicotine, arsenic, mercury, lead, and pyrethrum. Each of these poisons worked best for certain insects and all were applied in different ways.[43] Mechanical application of early insecticides was limited, and the case with grasshoppers provides a typical example. While much bait was spread by hand, mechanical spreaders were sometimes used, especially in the late 1930s when infestations were high in the southern parts of the Prairies. They saved time and were efficient because they spread the bait more thinly and evenly than was possible by hand. While a wide variety of spreaders was used on the Prairies, most were attachments for the rear of a wagon or truck. Usually the bait was dropped from an overhead hopper onto a spinning plate or disc which flung the bait out onto the field. Spreaders were sometimes powered by a small gasoline engine, but usually a hand crank or a chain drive operating from the wagon wheels was used.

Many of these bait spreaders were home-made, but a few companies manufactured them after World War II. The Calgary Farm Machinery and Supply Co., which also manufactured harrows and distributed other farm equipment, manufactured a bait spreader in 1950. Yet, as a field for manufacturing, bait spreaders were inherently limited. They were easy to construct at home, and the market was unstable because the equipment was not in demand once a cycle of infestation ended. In any event, the use of baits decreased because of the major developments in new types of pesticides in the late 1940s.[44] By 1945, DDT, "the miracle wartime insecticide," was available for public use. A chlorinated hydrocarbon, DDT was selective and was said to be nontoxic to mammals. By 1950, a huge range of other chlorinated insecticides had appeared on the market. Some of these were systemic, that is, they made the plant itself toxic to insects. While these poisons were available in dust forms, on the Prairies they were usually used in a liquid form sprayed onto the crop.[45]

Spraying

Before World War II, herbicides and pesticides, whether in liquid or dust forms, were primarily used for local application and were rarely applied to a whole field. In this environment, hand sprayers were usually used. After 1945, the general application of chemicals to whole fields demanded new types of equipment. Early field sprayers used low pressure and were mounted on a trailer, a tractor, or, less often, on a truck. All consisted of a tank, pump, and a boom fitted with spray nozzles. Tractor-mounted types sat saddle-bag fashion on the tractor. Their major drawback for prairie conditions was their relatively low capacity, and trailer types were commonest for field use. These sprayers came either with an independent engine to power the pump or were operated by a tractor power-take-off.[46]

Since farmers needed this equipment to use the new chemicals, its adoption was rapid. The case in Alberta was typical. In 1947, so few farmers owned sprayers that the province supplied the equipment or employed custom-spraying companies. Within a year, however, Alberta farmers owned 2,500 sprayers. Many of these sprayers were manufactured and sold by chemical companies and full line implement companies. There were, as well, a significant number manufactured on the Prairies. Versatile was among the early sprayer manufacturers, another instance of the company's good fortune in entering a manufacturing field on the eve of rapid change and growth. Among other early sprayer manufacturers was Golden Arrow Manufacturing Ltd., Calgary, established in 1947 by J.E. (Ned) Palmer. By the mid

1960s, Golden Arrow was the largest Canadian manufacturer of sprayers.[47] It manufactured complete spraying outfits, including aluminum tanks, nozzles, booms, and pumps. In 1970, George White and Sons Ltd. of London, Ontario, purchased the company and moved it to London.[48]

Golden Arrow and Versatile were only two sprayer manufacturers among many that appeared throughout the Prairies after the 1950s. While most offered different features on their sprayers, the general technological principles of sprayers remained largely unchanged. But sprayer size increased immensely, reflecting both the increased scale of farming and the greater use of chemicals. As well, the steady application of herbicides and pesticides created demand for more specialized sprayers for nonfarm uses such as golf courses, yards, and parks.[49] There were also ongoing developments to increase sprayer efficiency in general. Different crops and chemicals required different spraying pressure and volume to ensure effective and adequate penetration of the chemicals. Sprays also had to be applied at different times to match plant growth and insect life cycles.

By the late 1940s, control of weeds and insects was beginning to rely more on the use of chemicals. In this crop-spraying demonstration, the Golden Arrow Service, Calgary, tested one of its sprayers. The disc trailed at the side marks the sprayed section of the field to avoid overlap. (Harry Pollard Collection, P3688, PAA)

An especially important consideration was the uniform application of the chemicals. Ned Palmer of Golden Arrow noted that "a close knit and even spray pattern" was essential for effective application of field chemicals. He argued that a boom-type sprayer would do this best, as long as the boom was stable. If it swayed too much, the spray would be heavier at the end of the sway than in the middle. Part of the solution was operating the sprayer at low, steady speeds to prevent the booms from whipping around. As well, Palmer found that a boom of up to 25 feet could be stabilized with light rods, but a boom of any greater length needed castor wheels as well. Even with these features, maximum stable length was believed to be only 40 to 45 feet. Subsequent developments, however, disproved such views. Richardson Great Northern Manufacturing Ltd., Winnipeg, established in the early 1950s, was an innovator in supporting booms with wheels and was able to control even greater lengths in this manner.[50] Additional problems in securing uniform coverage related to control of sideways movement of the sprayer boom, especially on hilly land. Golden Arrow helped to minimize this problem through development of a self-levelling system for its sprayers.[51] These problems were also alleviated by the development of electronic monitoring systems which helped to regulate the spray to ensure correct application.

Other areas of concern were also addressed. Precise direction of the spray onto the crop had always been a priority since it ensured uniform application, saved costs, and lessened the potential harm of the spray drifting around the countryside. High air blasts were used to direct the spray more precisely, while cones were placed around the sprayer's nozzles to help direct the spray downwards. Shields were also placed along the boom to help prevent spray drift and to create better penetration of the spray onto the crop.[52] The introduction of a variety of marking systems addressed the problem of overlapping coverage. These used strips of paper, foams, or disc-type markers to cut a line in the soil so that the sprayed areas were easily identified. Ongoing maintenance affected sprayer efficiency since nozzles had to be kept clean and had to be replaced regularly to avoid distortion of the spray pattern and waste of expensive chemicals. A significant development in this respect was Golden Arrow's production in the early 1960s of clamp-on-type sprayer nozzles, which not only made them interchangeable but also easier to clean and replace.[53]

The extensive use of chemicals represented a significant change in postwar prairie agriculture, and its impact on manufacturing in the region was important. High demand for new, specialized equipment like sprayers caused rapid growth in the number of firms manufacturing sprayers. However, by the 1970s, many of these manufacturers had stopped producing sprayers or had gone out of business entirely as production outpaced the market.[54] Manufacturing picked up again in the 1980s when further efforts to improve production saw the spraying of additional chemicals such as growth regulators and desiccants. As well, the practice of spraying fallow fields, and while seeding, became increasingly widespread. Sprayers were mounted on seeding machines and herbicides were applied to the soil as it was being seeded. The sprayer tank was mounted on the tractor or towed by the seeding equipment. Moreover, a long-standing distinction between methods for applying fertilizers and those used for chemicals was becoming blurred. By the 1980s, chemicals in granular form (and fertilizers in liquid forms) had become relatively

Large field sprayers, such as this one manufactured by Brandt Industries, Regina, accommodated the larger farms and increased use of chemicals in prairie farming during the 1980s. The cones around the nozzles were designed to limit drift of the spray. (PAMI/Alberta Agriculture Photograph)

common. Coupled with the growth of pneumatic systems for seeding and application of chemicals, these changes in type and methods of applying chemicals marked a radical shift in agricultural practices.

Fertilizing Crops

Before World War II, enhancement of soil fertility relied almost exclusively on the use of green and livestock manures. While chemical fertilizers were available from the turn of the century, they found little use on the Prairies until the late 1940s. Without an understanding of the relationship between fertilizers and prairie soil, or soil surveys to determine the types of fertilizers needed, or the equipment to apply them easily, farmers were reluctant to use chemical fertilizers. Many experts during the 1920s also believed that chemical fertilizers were not useful on the Prairies. Within a decade, however, this opinion had changed completely when experimentation proved that chemical fertilizers could bring about higher yields on prairie soils.

After 1945, there was a dramatic increase in the use of chemical fertilizers on the Prairies as farmers recognized that fertilizer would substantially enhance production. Higher levels of use were also confirmed by the increased use of broadcast systems of seeding. Unlike seed drills, which placed the seed and fertilizer in narrow rows, the use of a one-way or a discer with a seed box required greater amounts of fertilizer because the seed was broadcast in wide bands under the soil.[55] By the mid 1950s, pressurized anhydrous ammonia had also come into use, but since its application required special equipment, it was not a commonly used fertilizer until the 1970s.[56]

The increased use of dry granular fertilizers after World War II required specialized equipment. The available fertilizers were not effective when broadcast on the surface of the field. They hindered germination if they came in direct contact with the seed and corroded the metal parts of seeding machinery. Because of this, conventional seeding equipment needed an attachment that would permit the seed and fertilizer to flow from separate boxes and place them apart in the soil. By 1930, Massey-Harris was manufacturing such attachments, and much developmental work was also being carried out by fertilizer companies. For example, Consolidated Smelter Products Ltd. (Cominco) manufactured a fertilizer attachment in Calgary in the late 1930s. Although this production ceased in about 1940, it was later resumed by Spierco Industries, Calgary. This company was formed in 1971 by Bill Spiers who had worked with Cominco in developing fertilizer attachments. Spierco built and marketed the Cominco fertilizer attachment and Cominco provided some financial assistance. In 1972, the product was selling well as a fertilizer attachment and was also finding additional uses in applying granular herbicides and seeding small seed crops such as rape. The company was also working on developing a model suitable for row crops.[57]

By the 1970s, pneumatic systems for applying fertilizer had also become common. While air seeders could be used to apply fertilizer, there was some demand for fertilizer applicators that worked alone or could be attached to other equipment. A number of manufacturers therefore applied the technology developed in pneumatic seeding to fertilizer applicators. In some instances, these approaches could be used for liquid fertilizers as well, although these fertilizers were usually injected into the soil. This was done either with a cultivator fitted with a distribution system or with spoke wheel injectors such as those invented and manufactured by Pattison Bros. Agro of Lemberg, Saskatchewan. Similarly, the use of anhydrous ammonia became more popular in the 1970s. It too required injection into the soil, giving rise to some manufacturing activity in supplying parts for this equipment.

The combined application of seed, fertilizer, herbicides, and pesticides that was possible by the 1970s went a long way to meet the prairie farmer's dream of performing all field operations without hired labour and with fewer pieces of equipment. Seeding, cultivating, fertilizing, and weed control were also coming together in a single operation in a historically unprecedented manner. These developments became yet more technologically complex with the use of air seeders. As technology related directly to regional conditions and needs, the air seeder's use and development on the Prairies formed a direct and often intimate complement to changes in tillage practices, particularly the commitment to retaining trash cover. And in the same manner, it was not surprising that prairie manufacturers were able to develop expertise in filling these specialized needs.

Harvesting Equipment

Harvesting was the focus of significant technological change, and its mechanization involved devising machinery for some of the most complex farm tasks. The revolutionary change from the flail to the combine harvester occurred within the space of less than a century. Like most other field tasks, harvesting was mechanized on the Prairies as early as the 1880s. O.S. Longman, who grew up on a Manitoba farm in the late nineteenth century, never saw a flail in use for separating grain from the head; even then, all grain was threshed with machinery. The other traditional harvesting equipment—the cradle, scythe, and sickle—were equally anachronistic by this time, although the scythe and sickle continued in limited use almost until the present.[1]

Despite this use of machinery, harvesting by the turn of the century was still labour intensive, involving the distinct steps of reaping, stooking, and threshing. Still, this represented a major advance over pre-mechanized conditions, and greater changes occurred in the twentieth century with the development of combines and swathers. The magnitude of the change is evident from the observation that cradling two acres was estimated to be a full day's work in the mid nineteenth century. But in the next stage of technological development, a binder could cut and bind 10 to 12 acres per day. In both cases, additional time was required to thresh the grain. By 1947, combining two acres with a 16-foot self-propelled combine took under an hour.[2]

Like wandering troubadours, threshing crews such as the Walker threshing outfit in 1895 moved throughout the rural areas during harvest. (Ernest Brown Collection, B268, PAA)

Binders and Stooking Equipment

One of the most significant events in the mechanization of harvesting in the early nineteenth century was Cyrus Hall McCormick's development of the reaper. It cut the grain and a revolving reel knocked the cut grain back onto a platform. A man walking alongside raked the cut grain onto the ground where it was bound by hand into sheaves. By 1876, modifications had transformed reapers into binders that cut the grain and, with an automatic knotter, bound it with twine into sheaves. This represented an important saving in labour. The sheaves were then set up by hand (or stooked) as bundles (shocks, or, more commonly, stooks) in the field and left to dry. Much of this grain was cut when still slightly green and so needed to dry before threshing—indeed, a common theory held that grain benefited from standing in shocks where it could "sweat," or cure. Once dry, they were gathered up and threshed. Stooks provided some protection for the unthreshed grain and they could be left in the field over the winter if necessary and threshed in the spring.

Cutting grain with a binder at Okotoks, Alberta, about 1918. The sheaves lying on the stubble would then be stooked by hand. (Harry Pollard Collection, P495, PAA)

By the 1850s, reapers were common in southern Ontario and in the midwestern United States. A few were also in use at Red River, and they represented another aspect of the mechanized agriculture that occurred on the Prairies at the same time as settlement. The first binder was used in Manitoba in about 1882, and hundreds were being shipped to the Prairies by the late 1880s. They were highly prized because they saved immense amounts of labour and made the harvesting of the large prairie fields easier and faster in the short period before the fall rains and the onset of winter. In 1887, the Massey Company shipped 87 binders to Brandon, and the town welcomed them with a carnival including a parade and speeches.[3]

Binders were produced by many companies and competition was fierce, although the largest of the companies amalgamated into the International Harvester

Company (IHC) in 1902. Like the Massey machines shipped to Brandon, all binders were imported to the Prairies. Prairie inventors and manufacturers largely concentrated on attachments such as mechanical stookers. By the early 1930s, many types of mechanical stookers had been built, and large sums of money had been spent in attempting to perfect them. Because stooking was such a labour intensive task, mechanical stookers were probably thought about as soon as the first binder appeared in farmers' fields.[4] Traditionally, stooks were not bound, and the sheaves simply rested one on another. However, if the stook was constructed too loosely, rain penetrated it, but if it was set up too densely, air could not move through the sheaves and dry the grain. In either case, the grain might rot. In short, it was hoped that the mechanization of stooking would lead to greater economy as well as efficiency.

The farmer's dream of being free from hired labour contributed significantly to the continual efforts in Canada and the United States to develop a mechanical stooker. Though the farm press continually announced that a workable automatic stooker had at last been invented, none were ever successful.[5] Yet numerous inventions were continually, and with extraordinary perseverance, brought onto the market. Models by the Western Stooker Co., Brandon, in 1908 and the Grand View Stooking Machine Co. of Grand View, Manitoba, in 1909, IHC's 1918 model, John Christie's prototype machine at Vulcan, Alberta, in 1925, and that of the Boychuk Stooker Patent Co. of Edmonton in 1928 were all examples of these efforts.[6] With this much activity, and the customary search by manufacturers for financing, it was

A threshing crew at Brandon, Manitoba, in 1900. Since the threshing machine did not have a blower, the straw was removed from the machine by hand. Later, blowers on the threshing machine automatically blew the straw into a stack, saving much labour. (Ernest Brown Collection, B691, PAA)

only a slight exaggeration to say in 1918 that anyone talking about a stooker on the Prairies was "apt to be misunderstood; he is likely to be mistaken for the advance agent of an outfit with stock to sell and an imaginary shocker [automatic stooker] that he claims to be on the point of beginning to manufacture." [7]

Although there was immense variety in these machines, most were a binder attachment in which the bound sheaves were gathered, tied with twine into a stook, and then automatically deposited on the ground as an upright stook. All had a variety of problems. They created a side-draught on the binder, making its operation "hard on the horses and hard on the binder." On rough or even rolling land, they would not form a stook that would stand up or contain the needed number of bundles. In other cases, they did not work with grain that was even slightly lodged, and other crop variations caused problems that were equally difficult to solve.[8]

Mechanical stookers promised to reduce labour demands on farms. Many were invented, but few worked well. This photograph shows the Boychuck mechanical stooker, Edmonton, in 1928. (ND–3–4276, GAI)

Hopes for a successful stooking machine were only one aspect of the desire for mechanized threshing operations. Handling sheaves during threshing also required a great deal of labour. There were various companies manufacturing sheaf loaders, which reduced the number of hired men and teams needed to pick up sheaves in the field and take them to the threshing site. All of these machines operated in approximately the same way by gathering the sheaves in the field with a pickup and elevating them into a wagon.

Among the most successful prairie companies manufacturing these products was the Stewart Sheaf Loader Co. Ltd., incorporated in Winnipeg but operating in Regina from about 1910 until at least 1919. Such machines were especially popular during World War I when labour shortages were acute, and in 1915, there were about 1,200 Stewart sheaf loaders operating in the Prairies and the northern United States. Some companies, such as Davidson and Herriot, and Acme Manufacturing, both of Winnipeg, made only the pickup and elevator. Others, like the Stewart Company (under the brand name New Era) and the Jackson Combination Sheaf Loader and Carrier of Brandon (later of Saskatoon), added a barge that automatically unloaded the collected sheaves. This system, however, was unwieldy because the weight was too heavy for the horses. In about 1928, Acme brought out a self-propelled loader, which basically was a loader attached to a Hart-Parr tractor. As the company advertised, the loader could be detached from the tractor, making the power unit useable for more than one task. Another significant change was that the system was equipped with lights and could be "used all night."[9]

By the late 1920s, stooking machines and sheaf loaders were fast becoming out-dated. Swathers were being developed and they, along with changes in combines, were radically changing harvesting. While stooking soon became a thing of the past, binders were not immediately replaced. Binders were used until after World War II, even on farms with combines and swathers. Some farmers would use a swather for wheat and then use a binder to save the straw in a better form for feeding and to harvest oats for feeding in the sheaf to livestock. In 1947, binding was still commonly used for sweet clover, alfalfa, and rye—all of which shelled easily when ripe—as well as for malting barley. Binders ultimately disappeared; adjusting reel speed in swathers (or when straight combining) was found to eliminate many of the problems of shattering, and other developments made binders and threshers obsolete by the early 1950s.[10]

From Threshing Machines to Combines

In the late nineteenth century, the flail was replaced with hand-operated threshers with a rotating cylinder that threshed the grain. Human power was in turn replaced by horses working a treadmill, which then gave way to steam power. Threshing was continually being speeded up, although it still took most of the months from August to October. Nevertheless, a man with a flail, working very hard, might have been able to thresh 10 bushels a day. The capacity of early hand-operated threshing machines was advertised as being upwards of 300 bushels a day, and the large threshing machines of the early twentieth century could deliver vastly greater output.[11]

By the beginning of the twentieth century, threshing machines (often called separators) had become huge, complex machines powered from belts run off steam and, later, gasoline traction engines. Large crews of up to 30 men, and as many horses, were required to complete the various tasks of running the engine, pitching sheaves into the machine, and removing the grain. In 1898, there were about 370 threshing machines in all of the North-West Territories, but by 1908, Saskatchewan alone had about 3,200. Sales continued to be high until the mid 1930s when combine sales began to take the lead. By 1945, only 88 threshing machines were sold in the three Prairie Provinces. Unlike the combines that succeeded them, farmers rarely purchased their own threshing machines because of their high purchase and operating costs. Small farmers either formed partnerships to purchase, maintain, and operate a threshing machine, or they hired custom threshing crews or the rigs of large farmers in the district.[12] These huge threshing machines moved continuously throughout the Prairies, and provincial departments of agriculture often assisted by giving farmers advice on appropriate rates as well as by co-ordinating applications for work by custom threshing crews.[13]

Few threshing machines were manufactured on the Prairies. Prairie manufacturers were the servants of the larger Canadian and American manufacturers and produced only attachments or ancillary equipment. Of these, feeding mechanisms for automatically carrying the sheaves into the thresher and blower attachments for discharging the straw were especially important. Feeders reduced labour costs because they cut the twine on the sheaf, opened it, and fed it uniformly into the cylinder, and a number of companies manufactured these items. Typical of this equipment was the feeder produced by The Norris Manufacturing Co., Winnipeg, in 1903. The company also manufactured sieves and blowers for threshing machines, all of them apparently manufactured under licence.[14] More innovative yet was the feeder manufactured by the Nelson Jackson Co., Saskatoon, in 1923. Its machine had the loader at ground level (normally it was about six feet in the air to receive bundles pitched from a rack). Despite this novel system, the company was never competitive. Blowers were initially used only on larger machines with sufficient power. They blew the straw away from the threshing machines and into a stack, which eliminated some labour. From 1917 until 1924, the A. Stanley Jones Blower Company, North Battleford, and later of Saskatoon, manufactured a blower designed by A. Stanley Jones. It was an important breakthrough because it worked on small threshing machines. It was designed to fit the "Call of the West" threshing machine, which was manufactured by Desjardins of Ste. Anne de Kamouraska, Quebec, and was distributed in the West by Stanley Jones.[15]

For most of the nineteenth century, there were a number of attempts to develop a combined harvester-thresher that would cut and thresh in a single operation. These efforts were unsuccessful until the 1880s, but combines were being used by 1890 in California. Shortly afterwards, they were also being used in Australia for harvesting wheat. Initially, they were designed to harvest a crop as it stood in the field, which meant that the crop could not be harvested until it was evenly ripe. Natural conditions favoured the use of combines in dry long-season places like California and Australia, but on the Canadian prairies, the season was usually too short for grain, especially hard spring wheat, to ripen uniformly. This was especially the case in the park-belt areas which had many low spots where growth was heavier and ripened later. Other problems were encountered as well. The crop could be damaged or even lost due to early frost or because of other perils such as sawflies which attacked wheat late in its growth cycle. Given these problems, the prairie adage was "cut on the green side if you want to save your crop." As well, weeds that remained green until the frost were common in most fields, and these jammed up the combine. In general, combines were unsuitable to prairie conditions and did not come into widespread use until after the development of swathers in the late 1920s.

The first recorded use of a combine on the Prairies was at Brandon in 1898—it was a Conroy's Combined Harvester and Thresher, manufactured in the Ottawa Valley. Another type of combine, which was essentially a mobile threshing machine, was the Sylvester Auto-Thresher, manufactured at Lindsay, Ontario, and sold in the Prairies around 1908. Two farmers at Spy Hill, Saskatchewan, imported a ground-driven horse-powered combine from California in 1908, and a few others were used in southern Alberta at about the same time.[16] There was, as well, some local invention. A "travelling thresher" invented by an (unnamed) individual at Calgary was demonstrated at the annual convention of the United Farmers of Alberta in 1916. The machine was apparently a conventional threshing machine equipped with an automobile motor. As the thresher moved along the field, sheaves were pitched into it to be threshed.[17] Inventions like this were widespread throughout North America and indicated the changing approaches to threshing that were in the air. While large companies led the way in marketing these inventions, it was not until the mid 1920s that the number of combines on the Prairies began to increase.

In 1922, Massey-Harris sent one of its self-propelled gasoline-powered combines to the Swift Current Experimental Farm for testing. To this point, the machine was not marketed on the Prairies because Massey believed that it was unsuitable for prairie conditions. However, the test results at Swift Current were so positive that the station purchased a machine. At about the same time, a few other combines were sold by IHC and J.I. Case to farmers in south-western Saskatchewan. This marked the beginning of combine use in western Canada. Sales increased "almost beyond practical bounds" after 1924. In 1925 there were 17 combines in use in the three Prairie provinces, by 1927 there were 791, by 1928 there were 4,448, and by 1930 there were over 9,500. As in other cases of mechanization, Saskatchewan farmers were the leaders in adopting the new technology. At first, they were used only on the open plains areas of western Saskatchewan (which had a longer growing season), but their use was spreading: by early 1928, Manitoba had just over 200 combines, Saskatchewan almost 2,400, and Alberta almost 1,100. In each of the provinces, they were used for all types of crops.[18]

These combines used mechanical principles similar to those on stationary threshing machines. A revolving cylinder and straw walkers separated the grain from the straw and chaff. The mobility of these combines, however, gave them a considerable cost advantage. Varying in width of cut from seven to twenty feet, combines were considerably cheaper to use than threshing machines, and narrow profit margins on wheat encouraged farmers to examine seriously all possible cost-saving measures.[19] More difficult economic conditions in the early 1930s, however, halted the spread of combines. As with tractors, sales fell drastically during the Depression. By 1931, sales in Saskatchewan had fallen to under 200.

In addition to poor economic conditions, a number of apprehensions about these early combines also hindered sales. In the early 1920s, some people felt that a complete change in harvesting equipment was too expensive given the experimental nature of the machinery. And, because these early machines did only straight combining, there was a good possibility that the entire crop could be lost because of hail, snow, or heavy wind, and "by shattering due to the necessity of allowing the crop to stand until it is hard." It was also rumoured that the grain trade might refuse, or at least discount, the price of combined grain out of fear that "it would not be properly ripened and hardened to make first quality flour."[20] By 1932, these fears had largely been overcome by public knowledge of prairie combine test results and by the possibility of modifying combines to harvest swathed grain.

Nevertheless, it was still popularly contended that combines were largely suitable only for dry areas where the crop could ripen fully before the September rains. If these conditions were met, and farmers adapted their farming practices to suit the combine and used "some judgement and good mechanical sense in its operation," it was thought to be "a profitable piece of farm equipment."[21]

Indeed, they proved profitable. As well as eliminating both the labour costs and the problems of using a binder and threshing machine and crew, combines also decreased the amount of crop lost during harvesting. The historian R. Bruce Shepard has calculated that because of these factors alone, combines reduced harvesting costs by almost 50 percent in the 1920s.[22] These cost savings were greatest on large farms because of economies of scale. Since combine use initially benefited larger farmers the most, this became another factor that contributed to larger farm size.

The general impact of the combine was similar to that of other examples of mechanization; it created a chain reaction that stimulated even more technological change. Trucks became more popular because threshing was now done in the field, and the grain was often trucked straight to the elevator instead of being stored and gradually marketed. This in turn created marketing problems since so much grain came on the market at once. More importantly, the need for more uniform ripeness (even with the use of a swather) brought changes. Evan Hardy noted that because of the combine, farmers began using "pure seed which would ripen early and not shatter excessively" rather than "the mixed varieties formerly used." Spring seeding also needed more careful attention to ensure that the entire crop matured at the same time. Weeds needed to be controlled better, and so did insects, which also contributed to uneven ripening. In 1933 it was noted that "anything that will diminish weed growth is perhaps more essential in combine farming than any other type of farming." Hence, good farming practices became even more important.[23]

The implications of such requirements were wide ranging. They favoured certain cultural practices and also brought changes to the farm schedule. Spring work became as vital as fall harvesting, which was now concentrated in a couple of weeks rather than the whole fall season. Thus,

> by eliminating the need for large crews to operate the binders and separators, and increasing the need to plant early so as to insure a ripe crop in time for combining, the combine altered the farmer's work year drastically. The combine also reinforced the trend towards power equipment. Not only did farmers now need tractors to pull the combines, they also saw the need for power to pull larger tillage and seeding implements in the spring.[24]

This increased acceptability of the combine was partly due to the advent of swathers, which, together with a pickup added to the combine, avoided the difficulties associated with straight combining on the Prairies. Although swathers increased the cost of harvesting, they "emboldened farmers to risk combines in districts that hitherto had been considered entirely unsuited to their use."[25]

The Contribution of Swathers and Header-Barges to Combining

Swathers and header-barges were significant elements in the move away from stooking and towards different methods of handling grain in threshing. The development of the swather meant that straight combining was now no longer the only way to use combines. Swathing could be done several days earlier than straight combining, a crucial lead in a region with a short growing season. Equally important, it allowed the combining of crops infested with weeds which "otherwise could not be harvested either until the weeds had ripened or until freeze up." Green weeds cut by a swather dried out quickly and caused "no further trouble in the harvesting operation."[26] With a swather, a crop could be cut at about the same time as with a binder, judged to be the stiff dough stage for wheat, barley, and oats.

The swather was largely a North Dakota and Saskatchewan development, and its creation was tied directly to early efforts to develop a combined reaper-harvester (that is, a combine). In 1908, Ole and August Hovland of North Dakota were granted a U.S. patent for a swather and a "travelling thresher." They had observed that loose, cut crop dried quickly when it lay on the top of the stubble. To capitalize on this observation, two machines had to be devised—one to cut the grain and leave it in a windrow on top of the stubble, and another one to pick up the grain when it was dry and thresh it on the spot. Following several years of experimentation, the Hovlands formed a joint stock company in 1909 and built prototypes of the two machines. Each machine was designed to operate separately. The swather was fixed on the front of a tractor, which pushed it along to cut the crop. It was made up of two sections, one on either side of the front of the tractor, each 15-feet wide with a cutting bar. The cut grain was carried to the centre by canvasses running on a table and was delivered through the centre between the wheels of the tractor and onto the stubble. The pickup on the travelling thresher was ground driven, but the machine had a motor to drive its threshing operation. If the crop was ripe enough to thresh after being cut, the travelling thresher could do so immediately; otherwise it was left to dry and then threshed in the field.[27]

These were remarkable machines. The combine was based on principles already in use, but the concept of employing it along with a swather was a radical rethinking of harvesting. But the Hovland experiments and designs drew little interest from either implement manufacturers or farmers. Although it would later be said that this apathy arose because "there was no interest, no faith in it," a broader range of factors was also at play. It was not enough to be inventive, even remarkably so. These machines not only posed a major challenge to established interests and practices, but they also represented so radical a change that a complete reorientation of thinking about harvesting had to occur before swathing could become a reality. By the mid 1920s, this reorientation had begun. Two factors were especially important in this changing attitude—combines, the other part of the Hovland experiment, were by then available, although only for straight combining, and there was a growing belief that threshing unstooked grain would be ideal.

In 1926, Ellert and Helmer Hanson, brothers farming at Lajord, near Regina, each built a swather. The Hansons knew of the Hovlands' experiments of 1908—they had grown up in that part of North Dakota and memory of the Hovland invention had "stuck in the mind." The main reason for the Hansons' efforts was the problem of getting harvest help. Available labour was "pretty awful, a few good men—but few." They hoped to reduce the cost of harvesting and "to establish a certain independence by requiring less help." In late 1925, they began thinking about the "possibility of picking up a swath and threshing it with a standard combine redesigned in part to handle such a row." Like the Hovlands before them, the Hansons were forced to develop a range of equipment. The crucial difference was that one-half of the needed equipment, the combine, while needing extensive modification, was already on the market and an increasingly accepted piece of equipment.

The swather designed by the Hansons formed the basis for swathers that would shortly be produced by full line companies. It featured a reel and a canvas belt running over a table that carried the cut grain into a windrow delivered through the centre of the swather. In 1927, an IHC engineer from Chicago visited the Hansons to see their swather. As Helmer Hanson recalled, "he looked smart and was smart"—his basic question was what would they do differently if they were to build another swather. They replied that they would make the rear opening larger for "heavy, tangled or weedy" material to clear. This advice was followed by IHC in their first swather, a centre-delivery unit directly based on the Hanson design. Helmer Hanson observed that when IHC "came into the picture we felt that two years

Swathers were essentially a prairie invention. IHC's first swather in the late 1920s was designed by the Hanson brothers of Saskatchewan. (IHC Windrow Harvester Operator's Manual, 1928)

of strenuous effort and awful chances taken with our crops had paid off."

The next step, that of adapting the combine, was a major undertaking and involved considerable risk because it was unknown whether or not the grain would keep in the swath. The Hansons initially modified the combine to pick up sheaves that "would cut the cost of threshing a great deal, but it was a compromise method if our row [swath] failed us." Pickups were not a new invention, having been used on sheaf loaders, but their application to a combine was new. J.I. Case specially tooled a combine cylinder to meet the Hansons' specifications, and the combine worked. After using it for picking up sheaves for two seasons, the Hansons changed to a straight swath "as it seemed to be a safe bet that the swath row would keep." The implement companies were now fully behind the concept.[28] They and the public were ready for the new equipment. By the 1928 harvest season, there were about 4,400 combines on the Prairies, almost half with pickups, and almost 2,000 swathers.[29] However, the Depression limited the adoption of combines with pick-ups, and it was not until the late 1940s that swathers came into general use.

While the swather confirmed the idea that unstooked grain could be handled efficiently, the concept was also being promoted on the Prairies by the use of the header barge (also called a header-stack barge) in the 1920s and 1930s. This system was used to cut and mechanically stack cereal crops to dry before they were threshed. Grain could be cut early because it could dry in the stack, and the header-barge also took less labour than binding and stooking. Most importantly, it was valuable in areas where crops were short and where a header, a push implement much like a binder, was used for cutting grain. Swathers were not suitable for short crops because swathed grain needed plenty of stubble to hold it up off the ground to dry.

In sum, binders and swathers were best for tall grains, such as hard spring wheat, while headers were best for short crops not easily bound into sheaves. One problem with cutting short crops with binders was that the reach of the reels was too high, and a popular solution was to nail strips of linoleum or tin on the slats to extend their pickup.[30] Even this was unworkable in many cases, and a header was used instead. The header cut off the heads of the standing crop and an elevator attached to the side of the header carried them into a wagon moving alongside. The grain was then pitched by hand into narrow stacks to dry. When it was dry, it was threshed with a threshing machine or a combine. If a threshing machine was used, the stacks were dragged to the threshing site using horses or a tractor. If the crop was ripe enough, it could be taken straight from being cut to the threshing machine, although two teams of horses were needed to do this efficiently. The combine, of course, could be driven to the stack where the grain was manually pitched into it.[31]

To eliminate the labour required to pitch a crop from the wagon onto stacks, around World War I a number of farmers in the Great Plains areas of the United States independently invented a barge. A wagonlike device, the cut grain was mechanically deposited into it and then dumped as a stack, reducing both labour costs and the cost of twine. At the end of World War I, these devices were also being invented by prairie farmers, especially in the dry southern areas where crops were

Harvesting swathed grain with a combine equipped with a pickup reduced the chance of crop losses because of poor fall weather. Photo taken near Lethbridge, Alberta, in 1948. (Harry Pollard Collection, P728, PAA)

often short. While header-barges were becoming obsolete in the U.S. by the late 1920s because of the adoption of straight combining, they provided a solution to the harvesting problems of the prairie west. Because binders were generally unworkable with short crops, farmers instead "turned to home-made header stacker-barges, which handled loose spikes with short straw and concentrated them in stacks for threshing." [32]

The drought of the 1930s brought about even greater use of this technology. Given short and thin crops, and unable to afford combines, Canadian prairie farmers (largely although not exclusively on the southern plains) adapted binders for use with a barge. These adaptations mainly involved removing the knotting mechanism and fitting the binder with a header attachment. An elevator, often home-made, or a commercial sheaf loader, was attached to one side to carry the cut grain onto the barge, and from there it could automatically be formed into stacks. Thousands of

Jackson Sheaf Loaders en route to customers. (From the collection of the Western Development Museum, Saskatoon, 6–C–13)

these barges were used in Saskatchewan and Alberta, and most were home-made.[33] Several types were popular, and the farm press, the experimental farms, provincial departments of agriculture, and the prairie universities helped educate farmers by providing designs and by building prototypes of barges made from recycled wood and other discarded material. It was an example of co-operation between farmer-inventors and scientists.[34]

By the 1940s, header-barges had disappeared. Farmers turned to combines as soon as they were able to do so, but as Thomas Isern has observed, the header-barge remains an example of a "broad and common environmental influence" on farm technology on the Great Plains area of North America. While noting that some header-barges were commercially manufactured, he estimates that only about 1 percent of those used in Canada were commercial models.[35] Full line companies showed little interest in manufacturing this equipment, although John Deere apparently did some developmental work on a model in 1923.[36] Instead, some prairie firms manufactured for this regionally specialized harvesting technique. One of these was the Best Way Harvester, which in the United States was called the New Way Harvester. Both firms were taken over in about 1931 by Gregg Manufacturing Co. of Winnipeg, which began as a manufacturer of yokes and singletrees. Gregg manufactured a complete outfit including header, elevator, and stacker-barge. The barge was cylindrical and had a revolving floor "with packers placed around the side. The stack is revolved as it is formed. The grain is delivered into the centre of the barge from the header so that the stack will be firm in the centre and . . . shed the rain. The stacks are discharged off the rear of the machine in such a way that they are left upright and firm so they will resist wind and weather." In Manitoba and Saskatchewan, however, problems with the soundness of the stacks were experienced.[37]

The Gilbert stacker-barge was an effort to reduce labour costs and facilitate harvesting of short crops in the mid 1930s. (Harry Pollard Collection, P733, PAA)

Another firm, the Automatic Harvester Co., Calgary, manufactured the Gilbert Harvester-Stacker from at least 1934 to 1937. The firm was established by O.N. Gilbert in partnership with a Calgary businessman. Originally from Nebraska, Gilbert farmed near Rosebud, Alberta, and designed his harvester-stacker in 1933. He manufactured only the barge, but his company retrofitted binders for use with it for $35. The firm experienced great difficulty in finding capital and sold its machines only for cash. By 1938, the company was out of business.[38] Commercial manufacturing of stacker-barges failed largely because this product was only popular in hard times and because it could also be easily constructed by a farmer using second-hand materials.

Subsequent Development of Combines

While the earliest combines had been self-propelled, declining sales in the 1930s led to their adaptation for a power-take-off to eliminate duplication of large costly engines. Nonetheless, the trend in combines was to self-propelled models. Massey-Harris, using designs of an Australian inventor, brought out a self-propelled combine in 1938, but it was too large except for the biggest farms. In 1941, Massey brought out a lighter model (the No. 21) which "revolutionized grain harvesting" and became the model for subsequent developments in self-propelled combines.[39] By 1950, self-propelled types represented about 50 percent of Canadian combine sales. Though they had advantages in speed and handling, especially on irregularly shaped fields, they were more costly than pull-type combines and were condemned by many farm observers as needless duplication of large engines and a waste of capital. In 1957, it was estimated that farmers could save 58 cents per acre using a tow-type instead of a self-propelled combine. Farmers, nevertheless, eagerly purchased self-propelled models, and by the early 1980s, the justification for self-propelled combines reflected the changes in the scale of prairie farming. In 1980, it was noted that while pull-type combines offered lower depreciation and investment costs and were suitable for small farms, the advantages of self-propelled combines included greater manoeuvrability which lessened costs by reducing time spent on the job. And, as long as labour was available, they freed the tractor for other field work.[40]

By 1944 in the Elrose-Rosetown district of Saskatchewan, 88 percent of farmers were using combines, up from 58 percent in 1939. About half of these combines were less than five years old, demonstrating that, even with a restricted supply, mechanization advanced appreciably during the war years.[41] After World War II, the number of combines increased even more dramatically. While there were about 11,200 on Saskatchewan farms by 1941, this increased to almost 43,000 by 1951 and 68,000 by 1954. Although there were many binders and threshing machines still on farm inventories, few of them were in regular use.[42]

After the war, combines were the focus for continued expansion and innovation. CCIL was manufacturing a pull-type combine in 1969, but a more significant manufacturer was Versatile, which produced its first tow-type combine in 1964 and a self-propelled version in 1966. By this point, the company claimed to have 20 percent of the prairie combine market. In 1970, it brought out a self-propelled combine with full hydrostatic drive and a double set of straw walkers which the company claimed ensured more efficient separation of grain from the straw and chaff. In 1984, it brought out a trans-axial rotary tow-type combine, and by the late 1980s, New Holland (by then owned by Ford which had also acquired Versatile) brought out what it called the world's first rotary combine.[43]

Versatile's SP420 self-propelled combine, which came on the market in 1966, was a significant measure of the growth of prairie manufacturing. (VFEO–Ford New Holland Canada Ltd. Photograph)

These rotary combines were a significant alternative to conventional combine technology. There were different types of rotary combines, and although it was an old principle, it was not applied commercially to any extent until after World War II. These combines separated grain by centrifugal force rather than by shaking as did conventional combines. The straw was tumbled in a slatted drum to separate the grain and a blast of air removed the chaff.[44] In the 1950s, William and Fred Streich, two brothers farming near Clandeboye, Manitoba, developed their concept of using rotating drums instead of shakers to separate the grain from the straw. This system had higher capacity and could be operated at faster speeds than conventional machines. It also had lower grain loss. Heralded with some exaggeration as "the first major engineering change in combines since 1864," prototypes were built by John East Iron Works in Saskatoon and they performed successfully in the 1960s and early 1970s.[45] While the company, Western Roto Thresh Ltd., manufactured about 40 combines, it ultimately failed. Rotary combines were more expensive than conventional ones, and a small firm could not compete with the full line firms once they began producing rotary combines.[46]

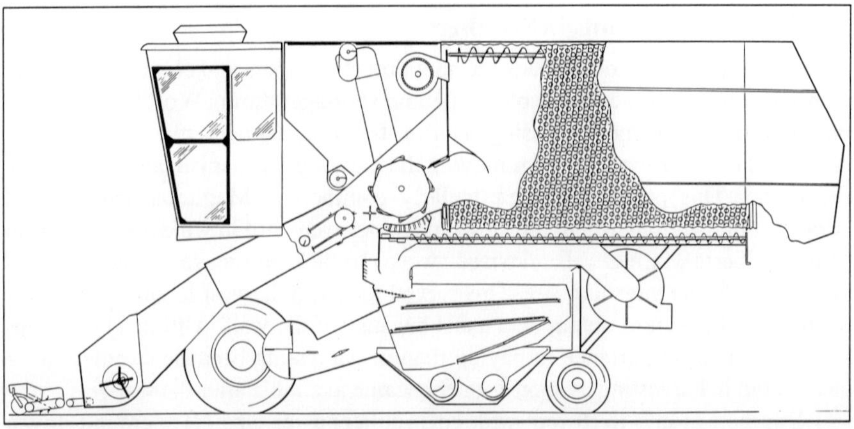

The Western Roto Thresh Ltd. combine was a prairie development that used a rotary principle in threshing, as this schematic drawing illustrates. (PAMI/Alberta Agriculture Photograph)

The demand for greater efficiency and speed, and the accompanying increase in the capital costs of harvesting equipment, created a need for various manufacturing enterprises. The growing use of pickup attachments on high speed combines created a highly specialized area for the manufacture of pickup reels, belts and tables, straw choppers, and chaff spreaders. New materials, such as nylon tines mounted in a flexible rubber base, represented the use of new materials for old functions. Other opportunities also arose in adapting combine parts to make them more efficient. As Jake Loewen of Altona, Manitoba, concluded in the late 1960s, since the combine's cylinder was crucial in the combine's efficiency, improving it would be highly beneficial and would justify expansion of his metal-working operation. In the late 1960s, he redesigned and manufactured a new style of cylinder bar, and he later expanded into a wide range of combine replacement parts which he manufactured through Loewen Manufacturing Co. Ltd.

Other manufacturing efforts concentrated on trying to reduce grain losses from the combine, which had traditionally been measured by counting the kernels on the ground of a harvested field. In the late 1960s, it was estimated that about 21 million bushels of grain were lost in Saskatchewan alone through combine inefficiency and improper operation. To help reduce these losses, the Department of Agricultural Engineering at the University of Saskatchewan designed and developed an attachment to measure losses. At first, the full line manufacturers of combines were not interested in this attachment, apparently because they saw it as an admission that their combines were inefficient. But in 1968, Smith-Roles Ltd. of Saskatoon began manufacturing and distributing this grain-loss monitor, which measured the amount of grain lost because combines were being driven too fast or were overloaded. The operator could then take corrective action.[47]

Postwar Developments in Swathers

Although the Depression set back the adoption of swathers and combines, farmers were eager to use the new technology. Labour shortages during World War II made this ambition even more pressing. During the war, farmers still using binders converted them to swathers by removing the knotting mechanism and reinforcing the frame. This practice was particularly common in Manitoba and parts of Alberta.[48] In locations where threshing machines were still in wide use (such as in central Alberta), farmers also devised sweeps to pick up sheaves in the field and carry them to the threshing site. This served to reduce harvest labour, which was even more difficult to obtain than usual because of the war.[49] Ultimately, sweeps were of greater importance in haying than in harvesting because swathers were substituted in harvesting as soon as they became available after the war.

Swathers in an 8- to 16-foot-wide cut in either power-take-off or ground-driven models were available after World War II.[50] The late 1940s saw a great deal of developmental work on self-propelled swathers. K.S. Reimer and Sons of Ste. Anne, Manitoba, for example, manufactured a self-propelled swather in 1949, and by the early 1950s, self-propelled swathers had become an important product of short line companies in both the United States and Canada. Among the earliest prairie swather manufacturers was Killbery Industries of Winnipeg. In 1951, it brought out a self-propelled swather based on the patented designs of a Minnesota inventor.[51] By the mid 1960s, swathers made up the greater part of Killbery's production, most of it under contract to full line companies. In 1967, it manufactured 1,600 self-propelled and 1,100 pull-type swathers. But it faced a problem common to those who produced on contract for more than one full line company: each company wanted distinctive equipment. In 1967, the Royal Commission on Farm Machinery was told "that Killbery had introduced a new swather for Cockshutt this year, and were now under pressure from Case to bring out something new for that com-

Killbery Industries, Winnipeg, was one of the first prairie manufacturers of self-propelled swathers. Many were made on contract to full line companies, a practice continued by Killbery's successor, MacDon Industries. (Canadian Farm Implements, December 1952)

pany." Indeed, the Commission was told that relations "between Killbery and the main line companies they serve were becoming strained over the issue of product development." [52] When Killbery was taken over by MacDon Industries, the firm continued to manufacture swathers on contract to full line firms. By 1977, for example, it was manufacturing swathers for Massey-Ferguson.

Another early innovator in self-propelled swathers was Versatile, which put a version on the market in 1954. This was Versatile's first self-propelled implement, and like Killbery, it used Wisconsin air-cooled engines. Like other swathers, it also had variable reel speeds for different crops and conditions. But Versatile's machine differed from other swathers in some respects. Most notably, it was the only swather on the market with a conventional steering wheel. Other competitive swathers used a lever-actuated clutch-controlled drive system which was awkward and had many moving parts.[53] In 1966, Versatile also brought out a new model in which the canvasses moved from one end of the table to the other. This let a farmer have centre delivery for heavier crops, as well as end delivery for lighter crops, in which case the cut could be increased from the standard 20 feet to 40 feet. In 1968, Versatile incorporated hydrostatic drive on its swathers, a development made possible because "mass production had made hydraulic drives available at a more reasonable cost." In 1967, Versatile manufactured 6,200 swathers, and by the late 1960s, it held the largest market share for swathers in Canada.[54]

Versatile was another important early manufacturer of self-propelled swathers and was the first to use a steering wheel rather than levers. This photo shows the SP400, one of Versatile's later generation of swathers, introduced in 1968. (VFEO-Ford New Holland Canada Ltd. Photograph)

By the 1980s, the market for self-propelled and tow-type swathers was sufficiently large on the Prairies to sustain a number of additional companies. CCIL was a significant prairie swather manufacturer, and Honey Bee Manufacturing, Frontier, Saskatchewan, also began manufacturing swathers. Greg and Glenn

Honey, two farmer-inventors from Bracken, Saskatchewan, established the company in 1978 to manufacture rod weeders. In the early 1980s, they designed a new type of self-propelled swather as well as one that mounted on a two wheel drive tractor. Production of the second type began in 1984—manufacturing was contracted out while Honey Bee handled assembly and sales. In 1987, the company took the major step of setting up a plant at Frontier to manufacture the swathers. These swathers were a significant measure of the scale of prairie farming. The self-propelled types came in three models up to 75-feet wide and were capable of laying a single, double, or triple swath. The tractor-mounted swathers came in 42- and 50-foot widths and could lay a single swath or two swaths side by side or separately.[55]

While no better for use with short crops, self-propelled swathers had advantages over tow-types in speed, flexibility, and efficiency. Killbery's version, for example, was said to be able to "circle a post or tree closer than the grain can be sown."[56] Features like this were attractive to many farmers, and while self-propelled swathers were not immediately cost effective—they were about double the cost of a tow-type—they became more economical as agricultural practices changed. In 1961 it was noted that "the key to wider acceptance [of the self-propelled swather] has been the development of attachments" that adapted it to more than one field job.[57] Both Killbery's and Versatile's self-propelled swathers offered an option for converting the swather into a self-propelled sprayer. With the increased use of liquid herbicides and pesticides in prairie agriculture, this was an efficient combination of machinery uses and made the swather into "a fast, efficient, light and highly manoeuvrable spraying machine." However, for these machines to be practical, manufacturers had to build them more sturdily than before. Other features were introduced as well. For example, the use of universal cutting bars meant that knives did not have to be changed for different crops, such as hay. This development led many farmers to replace mowers with swathers.

Specialized Harvesting Equipment

In addition to grain crops, beans, peas, corn, potatoes, and sugar beets were also grown on the Prairies. Each of these crops required either special equipment or modification of standard grain-threshing equipment. Peas and beans, for example, could be harvested with a threshing machine by converting the cylinders and changing the speed to accommodate the larger size and brittleness of the seed. Combines made this harvesting even simpler. Although special sieves and screens were sometimes required, adjusting the clearance of the cylinder and operating it at a lower speed were usually adequate. This worked for peas and beans, and the adaptation needed for clover and small grass seed crops was little different.[58] Corn required special harvesting machinery such as picker-huskers, binders, and picker-shellers, all of which were available from an early date. Most of this equipment was manufactured by full line companies. Corn was largely grown in southern Manitoba and Alberta, and this market was too small either to sustain prairie manufacturers or to give them a base from which to compete with the full line companies or specialized American short line companies.[59]

Sugar beets and potatoes were also grown commercially in the southern parts of Manitoba and Alberta, and the sugar beet industry provided some opportunities for regional manufacturing. Around World War II, representatives of the Lethbridge Research Station travelled to Utah and Idaho to study machinery used there for harvesting sugar beets and potatoes. They returned with a proposal to construct a machine that would cut off the tops and harvest beets. Similarly, potato harvesters observed on the trip were deemed to be a product that "could be made in a good machine shop." Yet, the mechanization of beet harvesting in southern Alberta was slower than in the United States, even though available machinery was efficient and the scale of farming in southern Alberta made its use as cost effective as hand labour. In 1953, about 27 percent of the Alberta crop was harvested with machinery compared with 76 percent in Colorado.[60]

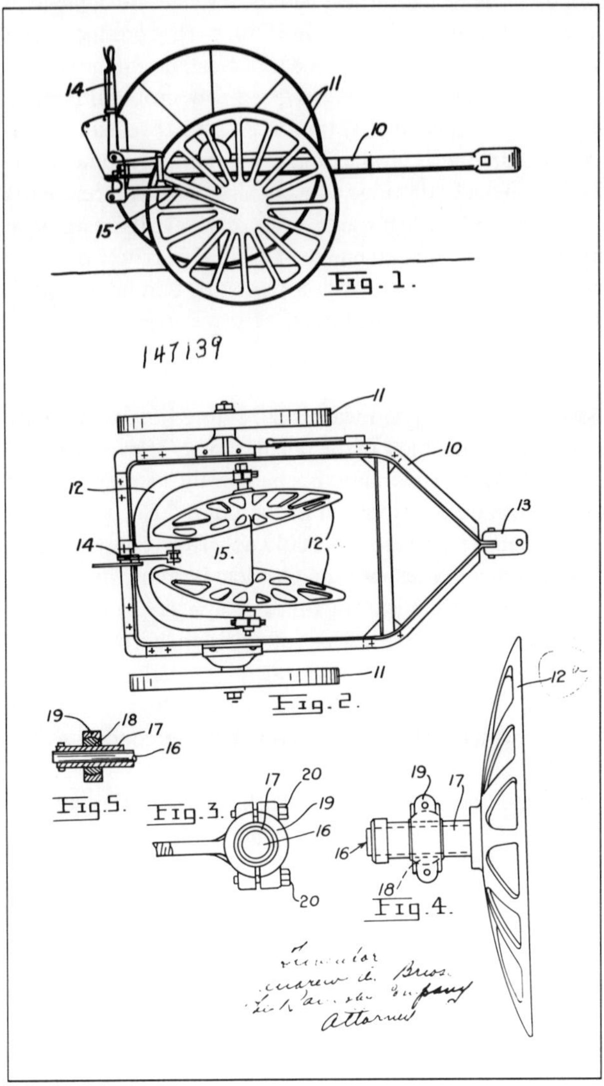

Beet harvesters were manufactured in Lethbridge, Alberta, by Kirchner Machine Ltd. They featured these lifter wheels designed by Andy Briosi, a local inventor. (The Archives, Sir Alexander Galt Museum, Lethbridge, P19851143002–GM)

Equipment for sugar beet growers performed a number of tasks, including lifting the beets, cutting off the tops, cleaning the dirt from the beets, and loading them into a truck. Some of this beet equipment was designed by Andrew Briosi, a brilliant farmer-inventor from

Picture Butte, Alberta. Briosi designed a press wheel lifter for harvesting sugar beets, which became the standard used for beet harvesters. Briosi also invented a beet thinner and weeder, as well as a topping machine. He manufactured some of this equipment himself, but in some cases, he found his patents infringed by U.S. manufacturers. In contrast, he was paid royalties on lifter wheels by Kirchner Machine Ltd. of Lethbridge.[61]

Kirchner Machine Ltd. was one of the earliest and most successful Canadian manufacturers of sugar beet equipment. Established in 1940, its first product was a hay sweep and stacker, but it expanded into sugar beet harvesters, thinners, disc toppers, and top savers. The company also manufactured row crop cultivators, which were used widely in sugar beet farming. By the 1970s, sugar beet equipment represented about 60 percent of Kirchner's production. Most of this equipment was sold directly to farmers, mainly in the Lethbridge area but also in the adjacent parts of the United States. This allowed the firm to provide direct service to its customers, and it also provided an opportunity for an interchange of design ideas between user and manufacturer.[62] Although American competition was stiff, Kirchner experienced little competition from Canadian manufacturers.[63]

Storage Equipment

Before the widespread use of combines, grain was usually stored for a relatively long period of time since harvesting took most of the fall and horse-drawn transport to elevators slowed marketing down considerably. Because of this, storage was of utmost importance to prevent loss or deterioration of the grain.[64] Even so, grain storage was often haphazard because many farmers could not afford commercially produced bins. Most built wooden granaries, or used obsolete farm buildings, and many also stored grain in piles on the ground during times of high yields. However, for those who had the income to buy them, a variety of steel bins were available from local manufacturers from an early date. These metal bins had a number of advantages. They were secure, they were vermin proof, and they often had ventilation devices on the roof which helped create good storage conditions for grain.

These bins were usually manufactured by metal-working companies such as the Jubilee Metal Cornice Works, Winnipeg, which specialized in culverts, metal roofing, barrels, and other metal products. Early metal bins were usually round and made of galvanized steel, and standard sizes included 8-, 10-, and 12-foot diameters and 8-foot heights. The bins came in sections which the farmer assembled on site. Most of these bins loaded from the side, although some had a manhole in the roof that could also be used for this purpose. In all cases, the peak of the roof featured a ventilation device. A few wooden versions, and an occasional square metal bin, departed from these typical forms. Regardless of their style, the fact that the bins were assembled on site led manufacturers to claim that they were "portable"; presumably, since the farmer assembled them, the farmer could also disassemble them. Thus, a popular feature was to limit the number of bolts needed to hold the bin together and make assembly faster. The Prairie Grain Tank Co., Winnipeg, for example, advertised in 1915 that one advantage of its bin was that it required less than 100 bolts, while competitive models required from 800 to 1,800.[65]

Bins like these remained typical until after World War II, and most continued to be manufactured by general metal-working companies until the late 1960s. Among the most successful of these was Westeel, which began in 1911 as the Winnipeg Ceiling and Roofing Co., Ltd., St. Boniface. Later known as Western Steel Products Ltd.—and still later as Westeel—it acquired a major metal bin manufacturer, Rosco Metal Products, in 1965. This created one of the largest metal fabrication firms in Canada. Constructing a wide range of metal products, it was a significant supplier of metal bins, which during the 1970s became typical on prairie farms. During this period, grain storage needs expanded, and the development of bulk fertilizers in the 1970s also created additional farm storage needs. By the late 1960s, the size of bins had increased dramatically. As was noted in 1968, while 1,200 bushel bins had been popular "a few years ago, 2,400 bushel models are becoming common today." [66] The trend in these bins was toward hopper bottoms. In addition to allowing faster unloading, aeration systems were devised to fit on the bottom of the bin to draw air into it for ventilation and drying. In some versions, an aeration device projected up inside the hopper bottom to increase the penetration of the air.

Many companies emerged in the 1960s and 1970s to meet this growing market. While the design of these bins

Among other products, metal bins were an early product of prairie metal-working companies. The Winnipeg Ceiling and Roofing Co. Ltd. was later amalgamated into Westeel Agricultural Products. (Nor'-West Farmer, June 5, 1915)

stemmed from the modular bins that had been typical until then, many of these more recent bins were welded units. As a result, they were a bulky product to ship, and this, perhaps along with increased demand, stimulated their manufacture at a number of points across the Prairies. Wheatland Bins of Lethbridge, for example, began manufacturing bins in 1980 and expanded in 1986 by opening a plant in Camrose, Alberta. This move was directly related to the need to be close to markets because of the bulkiness of the product. Metal hopper bins were also manufactured in the same district by other firms, including Chigwell Industries Ltd. of Lacombe. This practice, in which a number of manufacturers operated relatively close to each other, was repeated throughout the Prairies. All of these companies attempted to achieve a product mix by producing other equipment such as trailers and metal cladding, as well as bins and other metal buildings. Packages for converting traditional metal grain bins into hopper-bottom bins were also popular, and they were produced by many bin manufacturers as well as by metal-working shops in general. Metal-working shops also often manufactured related products such as bin ladders and bin-level indicators, continuing a long tradition in the history of prairie metal fabrication in which a firm manufactured a variety of products.

Grain-Handling Equipment

The handling of threshed grain took a great deal of labour, and mechanization of this task was an early target for inventors. By 1905, a drag-type elevator that moved the grain on a chain-driven conveyor was typically used for moving threshed grain. Many of these elevators were home-made affairs, although several companies manufactured them.[67] Most were stationary, but portable grain elevators were being manufactured in Winnipeg as early as 1903. However, they were still considered a novelty in 1917 when they were described as able to do "the work of two to five men."[68]

While the auger (that is, a screw within a tube) was used to move material in ancient irrigation systems and, more recently, in threshing machines, it was not used for moving threshed grain in or out of bins until after World War II. Elevator-type grain movers remained attractive until the 1950s because they could be used for bagged feed, ensilage, and most other products except loose hay.[69] They were, however, becoming out of date for moving grain threshed with large combines because they were neither fast nor efficient enough for the huge volumes of grain being harvested.

While one Manitoba farmer-inventor outlined the auger principle in 1923 to readers of *The Nor'-West Farmer,* it was not applied with precision to a manufactured product until World War II.[70] Essentially a prairie development, grain augers were not manufactured by the full line companies because they were too small an item. Without well established competitors, these augers therefore became an important new field of manufacturing for prairie short line firms.[71] Riding the crest of these changes was Peter Pakosh, who was among the first to make commercial application of the auger principle for moving grain. While still located in Toronto, Pakosh designed a grain auger, and since it could move grain up or down the tube,

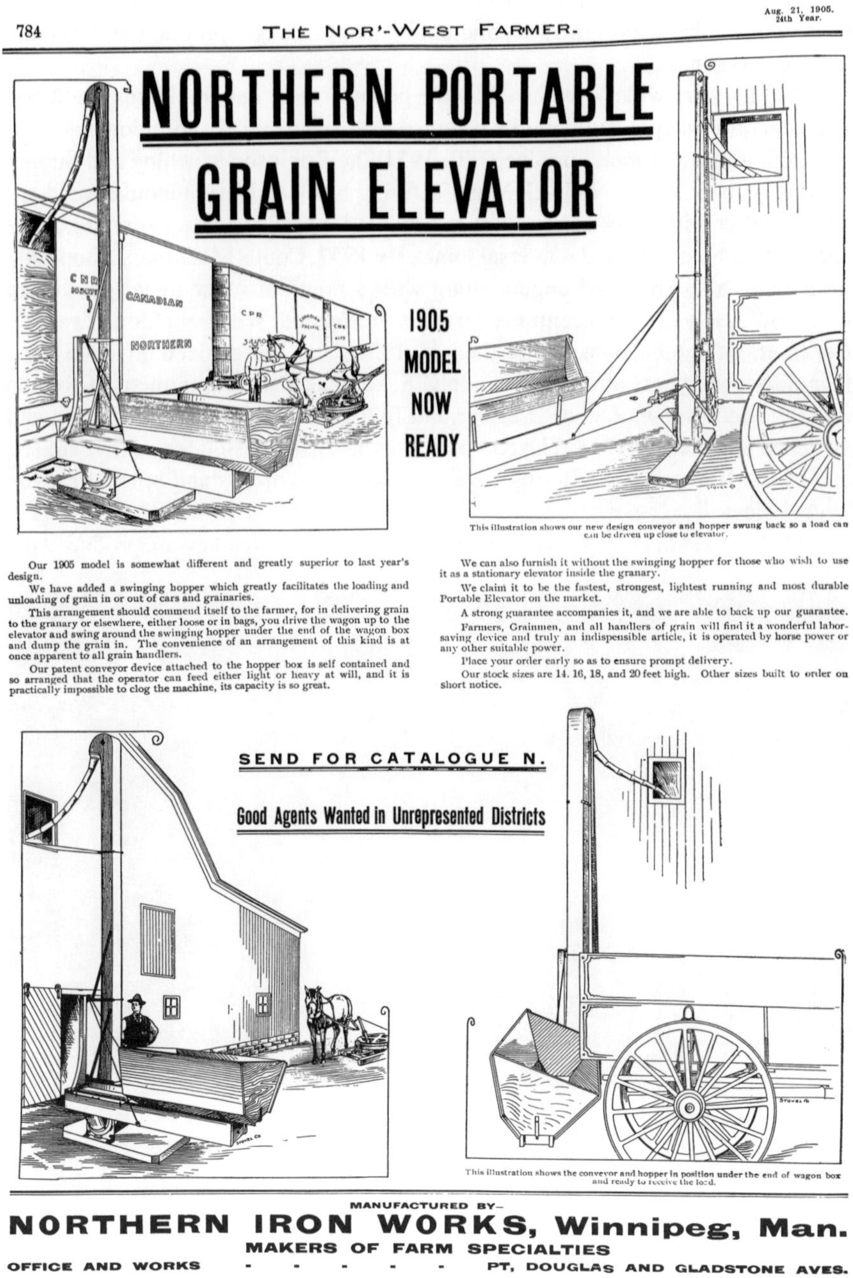

784 THE NOR'-WEST FARMER. Aug. 21, 1905. 24th Year.

NORTHERN PORTABLE GRAIN ELEVATOR

1905 MODEL NOW READY

This illustration shows our new design conveyor and hopper swung back so a load can can be driven up close to elevator.

Our 1905 model is somewhat different and greatly superior to last year's design.

We have added a swinging hopper which greatly facilitates the loading and unloading of grain in or out of cars and grainaries.

This arrangement should commend itself to the farmer, for in delivering grain to the granary or elsewhere, either loose or in bags, you drive the wagon up to the elevator and swing around the swinging hopper under the end of the wagon box and dump the grain in. The convenience of an arrangement of this kind is at once apparent to all grain handlers.

Our patent conveyor device attached to the hopper box is self contained and so arranged that the operator can feed either light or heavy at will, and it is practically impossible to clog the machine, its capacity is so great.

We can also furnish it without the swinging hopper for those who wish to use it as a stationary elevator inside the granary.

We claim it to be the fastest, strongest, lightest running and most durable Portable Elevator on the market.

A strong guarantee accompanies it, and we are able to back up our guarantee.

Farmers, Grainmen, and all handlers of grain will find it a wonderful labor-saving device and truly an indispensible article, it is operated by horse power or any other suitable power.

Place your order early so as to ensure prompt delivery.

Our stock sizes are 14. 16, 18, and 20 feet high. Other sizes built to order on short notice.

SEND FOR CATALOGUE N.

Good Agents Wanted in Unrepresented Districts

This illustration shows the conveyor and hopper in position under the end of wagon box and ready to receive the load.

MANUFACTURED BY—

NORTHERN IRON WORKS, Winnipeg, Man.

MAKERS OF FARM SPECIALTIES

OFFICE AND WORKS - - - - - PT. DOUGLAS AND GLADSTONE AVES.

At the turn of the century, grain was moved into bins or wagons by elevators, such as those manufactured by the Northern Iron Works, Winnipeg. (The Nor'-West Farmer, August 21, 1905) he noted that it was "versatile," which gave the new product its brand name. His auger was on the market in 1945, and until it stopped auger production, Versatile sold 50,000 augers.[72]

Augers were widely produced by the postwar short line implement industry. The general principle could not be patented, and augers nicely complemented the production of other metal products. By 1951, the Robinson Machine and Supply Company (later Robin-Nodwell Manufacturing Ltd.), Calgary, manufactured augers together with other agricultural and industrial products like hydraulic cylinders, power-take-offs, and universal joints. By 1947, Coutts Machinery Shop Ltd., Edmonton, also produced augers along with a range of other metal goods and services.[73] A few firms concentrated on auger production. Westfield Industries Ltd., Rosenort, Manitoba, was established in 1950 (and reorganized in 1965) and manufactured augers almost exclusively. Sakundiak Farm Equipment of Regina initially tried to manufacture a discer, but found it too large an undertaking and switched to augers in 1948. The company purchased the design for its first augers, and by the 1960s, it was offering augers in various diameters and in lengths from 29 to 49 feet.[74]

In about 1960, Peter Bergen Industries of Drake, Saskatchewan, produced the

Augers were an important prairie development after World War II. At the same time, hydraulics were becoming more common on farm equipment. In this 1958 photograph, grain is being unloaded by auger from a truck equipped by a hydraulic hoist on the farm of A. Fahlman, Kronau, Saskatchewan. (58–279–08, Agriculture Machinery, SABR)

first large augers in Saskatchewan, and other manufacturers began offering similar products in the 1970s. The increased size of augers was related directly to the use of larger bins. While early augers had been run off small gasoline engines, the large versions were adapted for use with a tractor power-take-off. Further innovation included the development of a side-delivery auger by Farm King of Morden, and yet others developed designs that allowed the auger to be lowered or raised from the tractor. Specialized augers also came on the market. Short augers for use under hopper bins or truck hoppers were manufactured by various firms. Some companies also manufactured models that modified the loading end of the auger by adding a piece of flighting (the revolving screw mechanism) that dragged grain from the bin into the main auger screw. Large scale farming created a need for yet another kind of auger—a small one with an adjustable long spout. Mounted on the rear of a truck, it was used for filling seed drills.[75]

The manufacture of augers demonstrated the changes in prairie manufacturing after World War II. In 1945, when Peter Pakosh began manufacturing augers, Brandt Industries of Regina also began producing them. Brandt made the flighting, as well as the tubing. The tube consisted of 4-foot sections welded together. This was characteristic of the earlier tradition of prairie short line manufacturing—in most cases, all fabrication was done in one shop—and there was little interdependence among companies. Some companies continued to produce all the components for their augers, but by the 1970s, this had changed and many were using components manufactured by others. For example, Uniflyte Co. Ltd., Winnipeg, established in 1975,

As the size of farms and the capacity of combines increased, larger augers and bins were needed. These augers, manufactured by Westfield Industries, are unloading into conventional steel bins. Hopper-bottom bins, such as the one on the left, were gaining in popularity in the 1980s and were manufactured at a number of points throughout the Prairies. (Westfield Industries Photograph)

manufactured flighting in addition to other specialized products for industrial and agricultural use. As well, another Winnipeg firm, Imperial Steel Products, manufactured steel tubing that was used by many auger manufacturers on the Prairies. The products of these various small manufacturers made it possible for some companies to incorporate specialized components made by others into a finished product, making this manufacturing as much a process of assembly as of fabrication.[76]

Besides the increasing variety of augers designed to meet specialized needs, the force of technological change found further expression in the area of moving grain. By the 1970s, blower systems that moved grain by pneumatic mechanisms were on the market. These were the same principles that had revolutionized seeding. Initially used for cleaning bins and truck boxes, equipment using these principles could also be used for moving grain from bins into trucks. These systems were advertised as capable of moving up to 4,000 bushels of wheat per hour, and several prairie firms were applying this technology in their product lines.

Drying Grain

Although grain dryers were used by the 1920s, they did not come into common use until the 1970s. In 1928, it was noted that there was "no method of grain drying adaptable to Western Canadian farm conditions that will give uniformly good results in all seasons." Putting grain through a blower had "little effect on the moisture content when it is above 16 percent." Indeed, it was recommended that tough grain should simply be marketed since "there will be greater weight by reason of the excess moisture; and the owner would be saved the labour and expense" of attempting to dry the grain.[77] At this point, few farmers used meters to measure the moisture content in their grain, and samples had to be taken to elevators. In the early 1950s, the Model 919 Moisture Meter was invented by a Mr. Greenwood of Winnipeg. It was subsequently developed by Halross Instruments (later Labtronics) of Winnipeg, which bought the rights to the meter in 1976. As a precise, portable moisture meter, it was an important breakthrough, and although it was used widely by grain companies and governments, its use by farmers was less widespread.[78]

By the end of World War II, grain drying was still viewed as "an expensive process."[79] Farmers who needed dryers tended to build them on the farm. Directions for construction were readily available, and one recommended approach was to modify a grain bin so that warm air could be blown into it. Although it meant less handling, this approach was not entirely satisfactory. High-quality buildings were needed, and in some versions, special architectural features had to be constructed.[80] The expense of these methods meant that grain dryers were still unusual on prairie farms by the 1950s, but there were ongoing efforts to develop systems that would be effective, portable, and affordable.

By the late 1960s, there were two basic types of grain dryers in use on the Prairies. Developing from the home-made systems of the 1940s and 1950s, there were "in-bin" types that were usually fixed inside the bin. These dryers dried the grain in batches: "the bin is filled, the heat is turned on for a period, and then the bin is emptied." The other type of dryer was the continuous-flow type (usually a mobile unit) that dried the grain as it circulated through the dryer. Both systems used heat produced by a propane heater while the fan was driven by a tractor power-take-off. Since temperature control was imprecise, grain dried by dryers was often suitable only for milling because its germination potential was sometimes damaged by excessive heat.[81]

The 1970s and 1980s were a period of increased manufacturing of grain-drying equipment on the Prairies. By the early 1970s, several grain dryers were in use. Some were fixed on the exterior of the bin at ground level. While this system had the benefit of low initial cost, the bottom layer of grain was sometimes overdried because of the placement of the dryer at the floor of the bin. Other dryers avoided this problem by using a screw-type auger to move the grain continuously around in the bin as it was being dried. These dryers were usually fitted inside a metal bin and required considerable labour to load and unload the grain. The preferable system was a more automated approach in which grain would be moved automatically through a dryer and into the bin. This favoured the use of continuous flow dryers. These dryers were powered by electricity or a tractor power-take-off, while the heat was produced by natural gas or propane.

A number of prairie firms produced drying equipment like this; one was Vertec Industries of Vermilion, Alberta, operated by George and Rosabelle Daugela from 1971 until the mid 1980s. Several years later, they established the International Business and Engineering Corporation (IBEC) at Spruce Grove, Alberta, and continued to manufacture a wide range of dryers. As only one further example, in the 1980s, Farm King Allied introduced (with its Buhler recirculating batch dryer) a portable drying system advertised as capable of drying up to 300 bushels per hour.[82]

Although the massive increases in energy costs after 1973 led to a temporary decrease in the use of dryers, they remained popular because they allowed harvesting of damp grain. Improvements in temperature control also ensured that the quality of the grain was not harmed by overdrying. According to Rosabelle Daugela of IBEC, these were strong selling points for dryers because they saved capital costs. A farmer could operate with one combine, harvest some grain when it was damp, dry it, and save the capital costs of buying two combines. In addition, it improved the grade of the grain and ensured higher germination.[83]

Harvesting, grain-handling, and storage equipment formed a significant part of prairie agricultural manufacturing, especially in smaller and less complex products like augers and bins. These two products alone have provided a substantial manufacturing base for many companies. Nonetheless, a considerable amount of harvesting equipment was also produced on the Prairies. While a number of combines were manufactured in the region, the Prairies were not, except for Western Roto Thresh Ltd., significant in innovation in this area. Combine technology tended to be developed elsewhere, and because it suited prairie needs, it did not inspire fundamental invention or adaptation. Prairie production of swathers, however, has been more widespread and longer lasting. This is understandable because the Prairies have been a major market for swathers. Essentially a Saskatchewan invention, swathers fit the pattern of innovation typical in prairie technological development. And while the first swathers were manufactured by full line companies, and were later produced for them on contract in the West, the extensive regional market for this relatively specialized product created opportunities for short line manufacturers to respond to local needs.

Hay and Livestock Equipment

Livestock traditionally took second place to grain production on the Prairies, although in some areas ranching and mixed farming created a more diversified economy. Poultry, although present on many farms, was a less well developed sector overall. By 1950, mixed farming was strongest around Calgary and Winnipeg. Dairying was carried out in all three provinces, but was particularly strong in the Winnipeg area. Manitoba, therefore, had the highest demand for mixed farming equipment such as grain grinders, manure spreaders, and silage equipment on the Prairies.[1] These patterns remained relatively constant in the following years. In general, however, equipment for livestock and poultry was a relatively marginal concern for prairie manufacturers. Since the same equipment that worked elsewhere in Canada met prairie needs adequately, regional manufacturers had little competitive advantage. Still, tasks like grinding grain for livestock and poultry feed and putting up hay did stimulate some manufacturing.

Ranching began in southern Alberta in 1881 and in south-western Saskatchewan in 1904, and although it came under increasing pressure from incoming farmers, it continued to be important in these areas.[2] In the late nineteenth century, cattle had been left on the range until they were four or five years old, but the terrible winter of 1886–87 killed large numbers of range cattle and ranchers began to reconsider this approach. A repeat of this disaster 20 years later finally convinced ranchers that supplemental feeding was necessary, and they began putting up hay and fencing sheltered areas for winter grazing. Another change occurred in the 1930s when the practice of selling 4- or 5-year-old steers began to decline in favour of the sale of those 12 to 14 months old.[3]

After World War I, farmers also began to produce more cattle to diversify their operations. Mixed farming had always been encouraged by government, and Saskatchewan, for example, assisted farmers in 1913 to finance the purchase of breeding stock.[4] Even so, field crop production still dominated on the Prairies. In Saskatchewan, for example, annual farm returns between 1908 and 1928 for livestock were between one-quarter and one-half of those for wheat.[5] Nonetheless, the number of cattle on the Prairies increased steadily, and in the century after the first ranches had been established, their number rose from under 1 million in 1900 to 7 million by the mid 1970s.[6]

By the mid 1950s, these cattle were increasingly raised in feedlots. While most feedlots were located in the park-belt region and in areas with irrigated lands, they also began to appear in dry-belt areas by the late 1950s.[7] At this point, beef cattle were primarily raised in south-east Manitoba, around Lake Manitoba, in south-western Saskatchewan, and in the foothills and central areas of Alberta. While the Prairies became a dominant beef-producing region in Canada, hog production on the Prairies was also significant, especially after World War II. Although hog production tended to increase in the traditional mixed farming areas in Manitoba and Alberta, there was also a substantial increase in numbers and scale of operation in central Saskatchewan after the 1970s.

Haying

Traditionally, hay was the most important farm feed crop on the Prairies, and it was produced everywhere in the region during the era when horses were used as a source of power. Until the 1940s, most of this hay was slough or prairie grass that was harvested from ditches, sloughs, and small, irregularly shaped fields. Before World War II, the most common system for putting up hay was to mow and rake it mechanically, and then to pitch it by hand onto a rack and then into a stack. Putting up hay was seen to be the most labour intensive task on the farm and was one of the reasons why straight grain farming was preferred to mixed farming—with its higher demands for labour and constant attention.

The use of mowers to cut grass and legumes for hay began at about the same time that other cutting operations (such as binding) were being mechanized. The first horse-drawn mowers were developed in the 1840s, although a satisfactory machine was not on the market until the 1860s. The design of these mowers provided the basis for mower technology for the next century.This static technology was dominated by full line companies and established short line firms outside the prairie region, although there were a few efforts to manufacture mowers on the Prairies. For example, the Canadian Farm Implement Co., Medicine Hat (whose affiliate, the Alberta Foundry and Machine Co. had manufactured "The Canadian" tractor in 1919–20), produced the "Wonder" mower in the early 1920s. Despite the company's claims, it seems to have represented little new in mower technology.[8]

Technological change in haying was instead concentrated on the gathering and stacking of mowed hay. Because hay had to be dried before storage, it was usually raked into a pile to cure. Hand raking could not keep up with mechanized mowing, and horse-powered rakes appeared shortly after the mower. By World War I, rakes were either dump or side-delivery types.[9] The dump rake consisted of a series of curved tines mounted on a wheeled frame. As the hay rolled up in front of the tines, they were tripped manually upwards to release a pile of hay. In this way, the hay was formed into small heaps across the field. Before World War II, these rakes were the most popular on the Prairies.

The other type of rake, with side delivery, swept the hay along its width and deposited it in a windrow. This type of rake had certain advantages. Hay in a windrow cured better than when left in piles by a dump rake, and it was also better suited for mechanized loading from the field. Since the loading of hay onto a rack required a great deal of labour, a number of hay loaders came onto the market. All were designed to pick up hay from the windrow and dump it onto an attached rack or wagon. These loaders were usually similar in principle to what was used for loading sheaves of grain onto wagons for threshing, and sometimes sheaf loaders were put into service to load hay. The disadvantage of those designed especially for loading hay was the fact that they only worked on level land and with domestic hay (such as timothy), which had evener length and consistency than wild grass. Because of this, they were rarely used on the Prairies, where slough and prairie grass were most often harvested for hay.[10] Instead, sweeps and stackers were more commonly used. They could often be constructed on the farm, their operation required less labour than did loaders, and they worked with hay gathered by a dump rake.

Sweeps were used to gather cut hay and carry it to the stack. They were in use in North America by the 1880s, and those made on the Prairies copied designs used elsewhere. Sweeps consisted of a frame with long teeth that "swept," or gathered, the hay in front of it. Since some sweeps could carry up to 600 pounds of hay per load, they were often equipped with wheels to stabilize the load. They were powered by horses or, later, tractors. A popular alternative to the sweep was a "bucking pole," a long pole (often fitted with teeth) with chains at both ends. A horse at either end dragged the pole along the field, gathering up the hay and dragging it to the stack. Since neither sweeps nor bucking poles could get the hay onto a stack, a separate piece of equipment, a stacker, was required for this task. There were many kinds of stackers; some used slings, others employed a fulcrum principle, and still others dragged the load of hay up a pole and dropped it onto the stack. Most of these stackers were home-made.[11]

Sweeps, such as the horse-powered one on the left, and stackers for lifting hay onto a stack helped reduce labour costs. (Harry Pollard Collection, P463, PAA)

By World War II, haying was still the "stepchild" in the family of farm crops as far as mechanization was concerned. Methods that had been in use for 60 years still prevailed, although new approaches were beginning to appear.[12] The war was a significant stimulus for this change because of labour shortages. In 1944, it was noted that "for either labour reasons or because of progress, people are realizing the benefits of mechanization of the haying operation."[13] Governments actively encouraged the use of sweeps and other mechanical devices for gathering stooked grain as well as hay to promote wartime agricultural diversification.[14] Even so, by the late 1940s, most prairie farmers were still using horse-powered mowers and dump rakes and were only beginning to purchase tractor-powered mowers, side-delivery rakes, and especially sweeps and stackers.[15]

Some of these field sweeps and stackers were produced on the Prairies.[16] Many were custom-made by blacksmiths, although there was some regular production as well. The most important development was a combined sweep and stacker, a single piece of equipment that attached to the front of a tractor. With it, the hay was gathered from the field, transported to the stack, and mechanically stacked. It paralleled front-end loaders which by then were gaining in popularity. These combined sweep-stackers concentrated all the weight of the load at the front of the tractor, and it was crucial that this weight be redistributed so that the sweep and tractor were balanced. One typical approach was to carry the weight with a wheeled frame that supported the sweep and projected in front of the tractor. In other cases, a superstructure was built above the tractor and was bolted to the tractor frame and to the rear wheels for stability.

Earl and Harold Kirchner, farmers in the Tilley area of Alberta, were early manufacturers of hay sweep-stacker equipment. In 1945, they patented a sweep for handling loose hay and manufactured 50 sweeps in their first year of operation. Production was then contracted to the Lethbridge Iron Works and also to the Alberta Foundry and Machine Co. in Medicine Hat. By 1950, the Kirchners had established a plant in Lethbridge to manufacture their sweeps, called the Ranch King Loader. Although the firm soon shifted its attention to irrigation and sugar beet equipment, their early work on sweeps was a significant measure of the technological change taking place in farm cartage.[17]

Sweep-stackers and similar haying methods were, however, gradually replaced with balers that packed the hay into rectangular bundles tied with wire, and, later, with twine. The earliest mass-produced balers were stationary machines and the hay was carried to them for baling. In the 1930s, pickup balers were developed that gathered up the hay from a windrow, packed and tied it into rectangular bales, and dropped them on the field. Yet, by the early 1940s, neither of these balers were common on the Prairies. Most farmers found them too costly given their limited haying needs. Baling also offered only a marginal saving of labour over the use of loose hay. While large ranchers in southern Saskatchewan and Alberta found baled hay more difficult to use, they were baling some hay for use in winter when stormy weather made the feeding of loose hay inconvenient.[18]

Despite these concerns, baled hay had a number of advantages. Overall, there was less loss in handling, it produced better quality hay, and baled hay took less storage room. It also meant easier handling for commercial hay producers who shipped hay by truck or rail to distant markets. As well, balers were of some use in gathering flax straw for sale for fine paper making, and in baling straw after it had been left in a windrow by the combine. There was a tradition on the Prairies of feeding straw, and in southwestern Saskatchewan and southern Alberta, it became "quite a valuable product" during periods of poor weather in the late 1940s.[19] Baling straw was a better solution than any other available method (such as straw bunchers that fit onto the combine), and this made balers popular in parts of the Prairies.

A combined sweep and stacker, which was operated hydraulically, was an important innovation because it saved labour in one of the most labour intensive tasks in farming. As shown in this photograph from about 1955, two Tisdale area farmers were able to carry out haying alone. (Wetherell Photograph)

Balers designed for use in eastern Canada and the United States were not wholly satisfactory on the Prairies where dust raised by baling was sometimes severe enough to damage the equipment.[20] Even so, prairie firms contributed little to the technological development of balers. CCIL carried out some research and development on a baler, but it encountered technical problems that it believed were not worth solving because of the nature of the market. By the mid 1950s, the leaders in baler production were IHC and New Holland, who were competing fiercely through price cutting. This, in combination with technical problems and tooling costs for baler production, convinced CCIL that manufacturing a baler was too risky and plans for its production were dropped.[21] By the end of the 1960s, balers still were not produced in western Canada. Instead, prairie innovation was focused on bale-handling equipment.

While baling had met a number of needs, the labour demands that had always characterized haying were still high. In the 1950s, bales were gathered on a rack pulled behind the baler. A worker on the rack stacked them as they were thrown onto the rack by the baler. This was hard work, and the bales then had to be unloaded at the stack or barn, although bale loaders using a chain drag (much like the early elevators used for grain) were used to make it easier. By the early 1970s, various sorts of bale-handling equipment had become available. Some attached to or were pulled by the baler and either gathered bales into bunches or pyramids, or stacked them on wagons. If bales were dropped one at a time on the field, equipment was available to pick up, haul, and stack them.

Haying at the Copithorne Ranch, Alberta, in 1949. Balers were at this point coming into use on the Prairies. The rake on the right is a dump rake. (Harry Pollard Collection, P395, PAA)

One of the earliest prairie firms to develop such equipment was Malmgren Manufacturing of St. Boniface, Manitoba. By at least the early 1960s, it had developed a unique device that attached to the side of a rack or a flatbed deck and picked up single bales from the field. As it moved along the row of bales, it grabbed the bales automatically, raised them up, and dumped them on the rack where they were manually stacked. This device was ground driven, and it could move about 100 bales in 20 minutes, which was much faster than they could be stacked by hand.[22] A more popular approach was to stack bales manually in the field into a pyramid consisting of 6 or 12 bales. They could be picked up later, ideally with a front-end loader. The bales were arranged in this pyramid, or stook, so that there was an air space at the bottom to prevent molding. In this form, bales could stay in the field for some time without spoiling.

This bale stooking was soon mechanized. George Williamson of Red Deer, who had long been involved in machinery design, devised a manually operated stooker in 1966 and an automatic version in 1970. Along with other equipment, he manufactured both types of stookers through Williamson Manufacturers Ltd., and they were immediately popular.[23] By the early 1970s, stooking had become the most common method of dealing with bales in the field. Agratech, Calgary, developed a fully automated bale stooker that built a six-bale stook, and other prairie firms were also manufacturing both manual and automatic types of bale stookers, as well as bale forks for use with a front-end loader.[24] The use of bale wagons also became

Bale conveyors, such as the Bradshaw Bale Booster, manufactured in Lethbridge in 1949 by the Lethbridge Iron Works, used the principle formerly used in farm grain elevators to move bales onto a stack. (NA 4220–8, GAI)

more common. In this system, balers were equipped with an attachment that tossed the bales into a large trailer towed behind. The bales were then dumped automatically under a shed for storage. Although it was expensive, this system was popular because it made baling into a one-person operation.[25] By the late 1970s, this, along with other changes, meant that stooking was becoming less and less common.

Even with mechanized handling, rectangular bales still demanded considerable labour, so they rapidly gave way when a better alternative was presented. In the 1970s, balers producing large round bales became available. The earliest of these were made in Germany. One type was imported by McKee Brothers, an Ontario manufacturer of haying and other farm equipment which had opened a branch in Red Deer in 1957. Operating under the direction of Jim McKee, the Red Deer branch imported a Welger round baler and had it tested by PAMI before introducing it on the market. It received an enthusiastic reception and sales were well above projections.[26] Round bales, usually with a diameter of six feet, shed water better and resulted in less spoilage. Equally important, their size required totally automated handling, which was now easily done with tractor-mounted front-end loaders.

The replacement of rectangular bales with round ones redirected development of bale-handling equipment. Small welding shops that produced a variety of trailers, and some large companies as well, now developed and manufactured large trailers for hauling round bales. Most of these bale haulers featured unique devices for unloading the bales. Other manufacturers produced special loading forks for

Round bales could be handled in various ways, but all were mechanized, a factor that helped increase their attraction. This bale spear produced by John Buhler Inc., Winnipeg, was designed to be self-levelling. (John Buhler Inc. Photograph)

lifting round bales with front-end loaders or mechanisms that attached to the rear of a half-ton truck to load one or two bales into the box.

Most prairie farmers now used baled hay, but forage harvesters were an alternative. Silage was an old form of feeding forage that was most popular in central Canada. But it was used to a limited extent in some large dairy and hog operations on the Prairies,[27] especially in Manitoba where corn and sunflowers were grown for livestock feed. Under this system, forage was cut and shredded while green and then stored in a silo. Silos were usually circular, tall structures built of wood or cement, although lined, covered pits could also be used. A few Manitoba firms manufactured wooden stave silos during World War I, and in the early 1920s, two large Saskatchewan lumber companies tried to promote construction of silos to increase lumber sales.[28] Silage was usually blown into a wagon and then into storage. This minimized handling but it required a high investment in machinery and buildings. Commercial hay producers could not use silage because it was too difficult to transport long distances.

On the Prairies, the adaptation of the principle of forage preparation for use with dry hay was more significant. McKee Brothers of Red Deer used the "shredder harvester" it manufactured to do this. This machine operated by picking up dry hay in the field and shredding it into an attached wagon. The chopped dry hay was then stored in a free-standing stack, which was found to weather well. The stack was formed either by using the blower mechanism of the shredder harvester or by dumping the chopped hay and stacking it with a front-end loader. It was a significant automation of haying and became popular in the 1970s.[29]

Care of Livestock

While hay made up the greatest part of livestock feed on the Prairies, it did not provide adequate nutrition and livestock were also fed grain, usually ground into "chop." This need was recognized at an early date, and the Economy Foundry Co. Ltd., Winnipeg, urged farmers in 1914 to buy its roller crusher mill since "numerous tests and experts" had shown that "stock do much better if fed on crushed grain than on whole."[30] Grinding grain was an ancient art—indeed, it was "among the first fruits of man's inventive genius." From ancient times, grain was either pounded, pressed, or rubbed into a finer form for human and animal consumption.[31]

On the Prairies, grinding equipment was used to prepare grain for livestock and poultry feed, and the ancient principles of grinding governed this activity. Grain grinders typically either pressed the grain (the edge runner or Chile mill), pounded it (hammer mills), or rubbed it (roller mills and burr mills). For farm use, only pounding or rubbing were significant and these methods were used in three types of mills. With hammer mills, the grain was ground by being struck by revolving hammers as it passed through the mill. This produced a feed of varied size as well as a great deal of fine powder. Roller mills used either smooth or corrugated rollers to crush the grain. They produced a uniform coarse feed, good for cattle and horses, but not for hogs. Burr mills were the least expensive of the grinders, passing the grain between two rough cast-iron plates, one of which rotated at high speed. All of these machines could be driven by a small motor, but usually power was supplied by a belt run off the tractor flywheel or, later, by a power-take-off.[32]

Grain grinders were an early product of prairie iron foundries. For example, Manitoba Engines, Brandon, was building grain grinders as early as 1908. Since the company also built small engines, it sold a roller-type grinder and an engine as a unit, although it sold them separately as well. During the interwar years, grain grinders continued to be produced by iron foundries and metal work shops.[33] Adolf Krushel of Morden, Manitoba, began manufacturing a grain crusher along with grain elevators and other conveying equipment in 1934. His company was called Standard Gas Engine Works Ltd. While Krushel was only one of many prairie manufacturers of grain grinders, he was among the most successful. Krushel was a farmer who had designed a burr-type grain grinder that operated by a belt run off the flywheel of a tractor, although later models also used a power-take-off. Because it incorporated some features not found on other crushers, it was very popular. Most were marketed through Macleods, a hardware chain store, under the brand name "Hi-Capacity." Between 1934 and 1970, about 90,000 of Krushel's grinders were sold, about 90 percent through Macleods.[34]

In many respects, Krushel was a typical—though particularly successful—prairie manufacturer of the 1930s and 1940s, and he exemplified the craft tradition of short line manufacturers. At the most, the company had about 40 employees. The war was a very profitable time for Krushel; farm equipment was hard to come by,

Adolf Krushel of Standard Gas Engine Works Ltd., Morden, Manitoba, is shown in this photograph beside a shipment of his grain grinders destined for Latin America. (John Buhler Inc. Photograph)

and prices were high. He used scrap metal, which he adapted or melted down, in order to maintain production. In later years, he added a hammer mill to his line, as well as other equipment like grain augers. But the business was always run without any formal organization. Krushel could barely read and write, and he kept no books. In the 1940s, sales sometimes reached as much as $250,000 per year, which was very high for a prairie firm at the time. But since there was little cost control in production or much reinvestment in the firm, by the late 1960s, the company was in trouble. Its manufacturing equipment was obsolete, and Krushel was then about eighty years of age. At this point, John Buhler took over the company.[35]

Buhler had worked for a short time for Krushel, but he had gone into other business, forming a car dealership in Morden. After buying Standard Gas Engines Works in 1970, he renamed it Farm King (then the brand name of the company's products). He reorganized the company and built a new plant, partly with the aid of a DREE grant. This was the start of a highly successful expansion for Buhler who showed a knack for turning around failing companies. In 1981, he bought the Winnipeg plant of Allied Farm Equipment Ltd., an American manufacturer of farm equipment that was in financial trouble. He followed this with the purchase of Standard Industries, a Winnipeg manufacturer of grain elevator equipment. These companies, as well as others controlled by Buhler, were then amalgamated into

Farm King Allied Inc., later renamed John Buhler Inc. By 1990, it had sales of about $21 million.[36]

Other Livestock Equipment

Compared with haying and grain-grinding equipment, barn, feeding, and watering equipment was a less significant part of prairie manufacturing. Before World War II, water was pumped with windmill power or by hand, and these needs stimulated considerable manufacturing activity in Manitoba from an early date. After World War II, electric motors for pumping water replaced hand or windmill pumps. By the 1930s, dugouts were commonly used as a water supply for livestock in parts of the Prairies, especially in Saskatchewan. By the 1960s, they were also being used to supply water for domestic purposes, but the water often became brackish in summer. It was soon found that pumping air into the dugout could minimize this problem, and among the innovators in the development of aerator systems was Koenders Manufacturing Co. of Englefeld, Saskatchewan.[37]

Steel or galvanized cast-iron water troughs, along with steel barrels, were an early product of iron foundries and metal-working shops. Water heaters were also popular. It was accepted wisdom that cattle were healthier and consumed less feed if they were given warm water. The earliest available means of heating water was to immerse a water heater in the stock trough to keep the water from freezing. These heaters consisted of a metal fire box with a chimney and an opening projecting above the tank to insert wood or coal. This device sat in the tank and kept the water warm. It was an awkward system, especially compared with immersible electric heaters, which were in use on the Prairies by the 1950s. In keeping with its higher number of mixed farms, Manitoba farmers were the leaders in adopting this sort of equipment, although other prairie farmers soon followed suit.[38]

A similar change took place in feeding equipment. Before the 1950s, self-feeders and waterers, which automatically supplied water and dropped feed into troughs, were not yet "extensively applied" in western Canada, although a number of home-made versions were in use on many prairie farms. Commercially produced types were largely manufactured by short line firms in eastern Canada and the United States.[39] Still some growth in manufacturing of this equipment did occur on the Prairies. Using the design of one of its founders, Westland Plastics Ltd., Winnipeg, manufactured hog feeders, among other equipment, but in general, manufacturing of this type tended to rely on technology developed elsewhere. For example, Edwards Grain Guard (formerly Edwards Rod Weeder), Lethbridge, expanded into livestock equipment in the early 1970s, manufacturing brand name equipment under license.[40] Additionally, prairie metal shops sometimes fabricated metal fencing, gates, pens, and floor grates, although significant growth did not occur in this area until demand increased after World War II. Further changes took place after the late 1960s when prairie firms picked up North American trends in applying plastics and fibreglass to such construction. Typically, livestock equipment was sold through farm supply stores and feed mills rather than farm equipment dealers.

Rural electrification as well as developments in conveying equipment, especially front-end loaders, also worked a major change in barn cleaning and manure spreading. Before World War II, most farmers manually cleaned manure from barns. This changed, however, and by the 1960s, the disposal of manure was becoming more highly automated. And with the growth of hog operations on the Prairies, automated systems came into wider use. For example, barn gutters equipped with a drag chain to remove manure had been used in early dairy barns, and they increasingly came into use in hog barns. A more typical method of manure disposal, however, was a liquid system in which manure was stored in cisterns and later pumped out and spread on the fields.[41] With the increased use of these systems, a number of manufacturers produced equipment to meet this market, but most of it drew more upon technological principles general to this sort of equipment than on unique prairie developments.

Although commercial manure spreaders were available from an early point, most farmers before World War II spread manure by hand from a stone boat. In Canada, the use of commercial manure spreaders seems to have been most common in dairy operations. Manure spreaders used a chain-driven conveyor belt that carried the manure to a rotating beater device (ground driven on the earliest models and power-take-off on later ones) located at the end of a wagon. As the manure reached the beaters, it was pulverized and flung out onto the field. The manure spreader was invented by a farmer at Stratford, Ontario, who sometime around World War I sold his business to IHC.[42] While manure spreaders were typically manufactured by full line companies before World War II, some prairie short line firms subsequently moved into this area of manufacturing. For example, Killbery Industries was manufacturing one for CCIL in the early 1970s.

Rural electrification also brought about substantial technological change in methods of barn ventilation. Before electrification, most barns were ventilated with architectural features or by attaching passive ventilators made of wood or tin on the roof of the building. Proper ventilation of barns was recognized as being essential for the health of animals and to prevent an excessive build-up of humidity. Roof ventilators were commonly made of wood or of metal. One metal type was made in 1918 by Metal Specialty Co., Regina, a manufacturer of grain-treating equipment. It was a revealing comment about the age that, while the company argued in its advertising that fresh air was essential to preserve the health of animals, it placed as much emphasis on the fact that its ventilators added "a wonderful touch of beauty to the building." It also pointed out that its ventilators suited "any kind of building and no wooden cupola can compare with them in looks or efficiency." [43]

Major change came with rural electrification, which allowed ventilation to be handled easily and effectively with electric motors. This brought about a significant shift in the look of livestock buildings and in animal husbandry practices. Architectural design became less important since problems like high humidity and varying temperature could be corrected with motorized fans. By the early 1950s, electric fans were coming into wider use on the Prairies in livestock buildings and in storage lofts where hay was kept.[44] It was not until the 1970s, however, that manufacturing of such equipment became significant on the Prairies. Airmaster Sales Ltd. of

In 1900, this wooden barn was equipped with a sling to lift hay into the loft. It also featured a wooden roof ventilator to aerate the loft where the hay was stored. These methods were replaced with electrical ventilation and feeding systems after the 1950s. (Robert Hoare Collection, H346, PAA)

Winnipeg adapted existing ventilation technology for a variety of agricultural purposes, and Wintech Inc. of Steinbach, Manitoba, manufactured a number of temperature control devices. Similarly, Dick van Ee and Larry Benning designed a heat exchanger for agricultural buildings and manufactured it, along with other barn equipment, through Del-Air Systems Ltd. of Humboldt, Saskatchewan.[45]

Technological development in livestock and haying equipment was comparatively late in comparison to developments in equipment for tillage, seeding, and harvesting. The range of equipment needed for this production on mixed farms had been relatively limited, but the advent of electricity created significant change in the sorts of equipment that could be used. In the economic history of the prairie region, livestock generally took second place to grain production, although by the 1960s, they were increasingly important. In technological terms, since prairie farmers found the existing equipment relatively satisfactory, only a small amount of basic innovation occurred. Only in haying equipment was there an appreciable degree of innovation—to lower labour requirements—and in many respects, these developments were part of the overall changes taking place in farm cartage.

Conclusion

The contribution of prairie inventors to the development of agricultural technology has been out of proportion to the region's population and economy. The list of prairie innovations is indeed impressive: the discer, various types of subsurface tillage implements, the swather, air seeders and floating-frame technology for heavy duty cultivators, rock pickers, and coil packers, as well as numerous smaller, but important, features for all types of equipment.

Before the 1930s, an immense standardization in farm equipment technology had emerged in North America. Had it not been for the unique conditions on the Prairies that forced innovation, it is likely that there would have been little change in most of the equipment used in the region. There had been numerous efforts and considerable talent expended in manufacturing equipment on the Prairies from the late 1880s, but few were successful. Most either produced goods with an uneven or small market share, or they could not meet the competition from well established short line or full line companies located elsewhere. Prairie farmers had not yet been forced by circumstance to seek alternatives to standardized North American equipment, but distinctive prairie conditions soon forced them to look for equipment that would respond to local needs. Often growing from inventions by farmers and blacksmiths, this equipment became the strength of the contemporary prairie farm equipment industry, which specialized in meeting needs not served by general North American farm technology.

Most of the standardized equipment in use on the Prairies was produced by full line companies which dominated and defined the farm implement industry. They profoundly shaped the broad standards and the pace of technological change. Moreover, their fully integrated system—a system in which design, manufacturing, distribution, and sales were intimately related elements—forced any challengers either to circumvent the system or to use a similar approach. The development of the prairie short line industry was significant because these full line practices, which had traditionally limited access by new firms to the North American implement market, were avoided because of necessity.

The development of farm equipment manufacturing on the Prairies was tied to the economic changes in prairie farming. Farmers' demands for labour-saving equipment and the steady growth of farm size were of direct significance to the full line manufacturers who both met and stimulated these needs at the same time. Their steam engines, threshing machines, binders, tractors, combines, and other complex equipment were basic in defining the general technology of farming. Within this general framework, however, short line manufacturers adapted and created equipment to meet specific local needs. This environmental explanation for the growth of the industry is revealed in the heavy emphasis on tillage and seeding equipment among prairie manufacturers. Although the regional market was a considerable one for small companies, it alone was not sufficient to accommodate the industrial structure of the full line companies. George Morris's unsuccessful attempt to interest them in his trip device for rod weeders was only one illustration of the emphasis placed by full line companies on general, not specific, technology.

Farm equipment manufacturing on the Prairies does not, however, reveal only the evolution and impact of technology. It is also a case study of regional development. This issue has been prominent among those interested in the history of the Prairies, and it has often been concluded that prairie economic diversification was held back by national economic policies and the nature of Canadian political life. The tariff, one of the central issues in this view, doubtless assisted Canadian full line manufacturers in central Canada. Yet there is scant evidence to conclude that without the tariff more extensive manufacturing of farm equipment would have taken place on the Prairies. Indeed, prairie short line manufacturers before 1944 rarely claimed that their business would expand with free trade, although some believed that the tariff gave them an advantage. While farmers occasionally argued that there would be increased manufacturing on the Prairies without the tariff, this rarely figured prominently in their concerns. Farmers hated the tariff because they believed it increased equipment prices, not because it might disadvantage prairie economic diversification. They were no more concerned with the fate of regional manufacturers than they were with central Canadian ones. Farmers' priorities were framed by what they saw as their self-interest, which was lower farm implement prices that were more in line with the income they received for their commodities.

When the tariff was reduced in the mid 1930s to a point where it no longer provided Canadian manufacturers with much protection, farm equipment prices did not fall. Nor did they fall when free trade was implemented after 1944. As a result, those critical of the tariff realized that their energy had been misplaced in the face of a market dominated by a handful of large firms. For some, this strengthened the argument that co-operative distribution and manufacturing were the means to lower implement prices. In fact, this approach was implemented through CCIL, but farmers as a group did not actively support the concept. Regardless of tariffs or free trade, most farm equipment continued to be high priced. Partly caused by the tendency towards monopoly in the full line industry, it was also the result of the industry's cyclical nature, the higher costs of manufacturing complex farm equipment, and other conditions peculiar to the farm equipment sector.

It might be argued that free trade after 1944 brought about the upswing in farm equipment manufacturing on the Prairies after World War II. But the sort of manufacturing characteristic of the postwar period, and the reasons for its growth, point instead to the importance of other causes. Generally, short line manufacturing on the Prairies after the war was a regional response to the unique needs of prairie agriculture in the wake of the environmental crisis of the interwar years. Because of this, equipment manufactured by short line firms on the Prairies usually did not compete directly with full line products. Versatile was the only prairie company to do so, and even then, it tailored its equipment (including its tractors) to large grain farms in the Canadian prairies and the United States great plains.

The role of freight rates in regional economic development must also be assessed in light of the nature of the North American farm equipment industry. Clearly, prairie farm equipment manufacturers were disadvantaged in comparison to central Canadian firms. Freight rates to ship farm equipment west were lower than those charged to ship the same goods east. In gaining national markets, the competitive advantage of eastern farm equipment firms over prairie firms was therefore obvious. Yet, crediting this advantage to discriminatory freight rates alone would be too limited. Central Canada's head start in manufacturing gave it an important advantage, which discriminatory freight rates confirmed but did not create. Massey-Harris, for example, was a large and well-capitalized firm with extensive export markets well before the Crow rate of 1897 or the lower freight rates on west-bound farm equipment that were introduced in 1919. In addition, there often was no substantial market in central or eastern Canada for the specialized goods (such as tillage equipment) produced on the Prairies. Even by the 1970s, the primary market for farm equipment manufactured on the Prairies was in the prairie region or in the dry-land parts of the United States, and not in central or eastern Canada. In other cases, goods manufactured on the Prairies similar to those manufactured in central Canada could not compete in the eastern Canadian markets because of shipping costs, whether subsidized or not. And given the Prairies' limited industrial infrastructure before World War II, prairie firms were at a further disadvantage.

While it might be countered that this limited infrastructure was one outcome of central Canada's tariff and freight rate advantages, this argument ignores the inherent advantage of older, settled areas over newly settled ones in respect to all of the factors that make up industrial infrastructure. This is not to argue that greater sensitivity to prairie industrial development in freight rates policy might not have been of at least limited benefit for prairie farm equipment manufacturers. However, other prairie sensitivities were of greater concern. Prairie farmers were firmly opposed to eliminating the East's freight rate advantage, and before World War II, they had enough political force to ensure that these rates were maintained.

The significant expansion in farm equipment manufacturing on the Prairies after World War II resulted in part from an expanding regional market. This prosperity shaped the prairie industry's success, and the extraordinary increase in consumption of farm equipment from 1946 to the early 1950s and during much of the 1960s and the 1970s created an unusually buoyant market, despite severe downturns in the mid 1950s, late 1960s, and late 1970s. The share of the farm equipment market enjoyed by short line firms increased substantially, although it remained a relatively small percentage of the total prairie market.

This general upswing was stimulated by changes in the infrastructure for prairie manufacturing and in methods of fabrication. Rural electrification made a contribution, as did the development of additional steel capacity. The increased use of component parts in manufacturing and the use of welded fabrication in farm equipment were both crucial changes. These changes all led to an increasing interdependency among those involved in the industry and a recognition of common interests. The maturing of the industry on the Prairies was best shown in 1970 by the formation of PIMA to work on behalf of the prairie industry. Government also became more directly involved in fostering industrial growth on the Prairies through encouraging, subsidizing, and stimulating economic development. While none of these factors alone accounted for the growth of the short line industry, together they explain its expansion. In sum, the development of the regional industry was thus the product of a range of interacting forces that created conditions favourable to manufacturing after World War II.

For the foreseeable future, the prairie short line industry will continue to be shaped by some of the forces that have conditioned the industry since the end of the late nineteenth century. It can be argued that, historically, financial success in farm equipment manufacturing has depended on bridging the hard times in a highly cyclical market. As well, many industry observers and leaders believe that the consolidation and mergers that have characterized the industry in North America will no doubt continue—successful small firms will be absorbed by intermediate or larger ones, and these in turn will be taken over by those at the top of the hierarchy. This results from the tendency to ever larger scales of production and the accumulation of large amounts of capital in relatively few hands—a basic element of a competitive industrial economy. But in the case of farm equipment, it is also stimulated by the cyclical nature of the industry, which increases capital needs and leads to instability that mergers and consolidations resolve, at least temporarily. This is unlikely to change since it is largely stimulated by the erratic nature of farm income, and, as many farm observers believe, a stable farm income seems unlikely to come for some time.[1]

Given the general inability to control their economic future, prairie farmers have tended to see technological innovation as an important solution to the problems they faced. The implications of this for the farm equipment industry are obvious. In addition to the innovation needed to improve existing equipment as field use demonstrates its weaknesses, ongoing innovation is also essential to preserve market share and meet more broadly based technological change. Technological change has been a part of the competitive structure of prairie farming for over three-quarters of a century and individual farmers continue to view it as necessary for their survival. Because of this, farmers will likely continue to demand greater sophistication in equipment, especially when their costs are high in relation to their income. In satisfying this demand, manufacturers will have to pay careful attention to the fundamental distinction between marketing and promotion on the one hand, and essential research and development on the other.

Both research and marketing have traditionally posed a problem for most prairie firms because of their small size. Continuing government and public agency support will remain significant in identifying markets and in responding to complex changes in agricultural practices. Although there has been some antagonism expressed within the industry to the conditions imposed by government in its public assistance programs, government insistence on accountability for expenditure of public funds cannot be faulted. This is not to argue that government cannot improve its administration of these programs. Regional criticism of the potential competition among the economic development programs of different jurisdictions in what is essentially a single market for farm equipment seems to be justified. Competition among programs, whether federal or provincial, would lead to irrational growth and increased capacity that could not be sustained in the region. In terms of prairie economic development, greater co-ordination and sensitivity in regional terms will be necessary for development of worthwhile and sustainable programs. The question is therefore not whether there will be an ongoing role for government in developing industrial strategies, but how that role can best suit long range regional needs.

The innovation of prairie manufacturers and inventors has often been a response to the demands created by the prairie environment. Clearly, many environmental problems are far from resolved. Wind and water erosion have recently increased, organic levels in prairie soils are declining, and increased salinity has emerged as a problem in some areas. As one scientist has concluded, "the principles of erosion control established 40 years ago are still valid today, but dramatic changes in mechanization and farming practises have apparently reduced the practical application of these principles." In other words, the methods developed for soil conservation, and the equipment that they required, may require re-evaluation.[2]

One example of this rethinking of agricultural field practices is the development of "zero till." This is a system that goes beyond the "minimum" trash tillage that has come to characterize prairie tillage techniques and moves towards a system where tillage is all but eliminated. Seed is planted "directly into the untilled stubble from the previous year using chemicals for weed control." The high-clearance hoe drill (such as the one developed by Charles Noble) and air seeders can be used in this approach, and, in general, less equipment is required overall. But farmer-inventors and prairie manufacturers will also have to design more specialized machinery for injection of chemicals and fertilizers into the soil. Proponents of this system argue that its economic and conservation advantages have been strengthened by the recent development of so-called nonresidual herbicides. Widespread adoption of this technique would have a significant impact on the industry. Manufacturers have so far largely been cautious, waiting for a clear trend to emerge before investing in the costly equipment design necessary for a changed approach. Seeding and tillage equipment would be most affected, but there would also be an impact on tractor use since the amount of field work would be reduced.[3]

Such new approaches are integral to the nature of competition in farm equipment manufacturing. While competition in the farm equipment industry has been fierce in the past, it has been within predictable limits. Full line manufacturers largely competed with each other through brand identification and technical innovation. Short line manufacturers have depended on specialization, quality, and, as the sector has matured, brand loyalty and reputation as well. Their small size has given them important advantages in lower costs of production and responsiveness to users' needs, and production of specialized products with limited markets will probably continue to characterize prairie industrial production.

At the same time, this advantage may be affected by ongoing changes in the nature of factory production. The recent dramatic growth in use of component parts in farm equipment manufacturing will likely continue, and while this development has benefited regional manufacturing, it may also threaten it. Simple assembly has no special or unique needs, so it could be moved to a location where it can be done more inexpensively. Methods of assembly line production of relatively complex goods that are tailored to small markets are already emerging. In short, the historically static quality of assembly line production, in which product uniformity has held the key to economical large scale production, may be changing. If these methods become established, they could mean the penetration of relatively small specialized markets (such as dry-land farm equipment) by large transnational firms. Their production capacity is flexible enough to let them combine their large scale research and fabrication advantages and assembly line methods with economical production of specialized lines.

While such future developments could fundamentally alter competition on a world level, most prairie short line companies have increasingly been concerned with gaining specialized export markets. They have tended to view the dry-land areas of the United States as a primary market. This is logical because of its proximity and its use of agricultural practices and technology similar to those used on the Canadian prairies. The fascination with exports cannot, however, cloud the fact that the prairie region has been, and will remain, basic to the security—and the ultimate prosperity—of prairie short line firms. They have, overall, manufactured for agricultural conditions that they know intimately, and this has made them responsive and innovative. Prairie manufacturers have realized that until equipment has been developed, accepted, and modified for its own market, it will not gain substantial and long-term favour elsewhere. As many industry leaders observe, recognition of this "regional" specialization has been central to the growth and success of the prairie farm equipment industry, and in a time of growing demand for specialized agricultural technology, it will continue to be significant.

Notes

Abbreviations

CFIEI: Canadian Farm and Industrial Equipment Institute
GAI: Glenbow Alberta Institute Archives and Library, Calgary
LRS: Lethbridge Research Station, Library
NAC: National Archives of Canada, Ottawa
PAA: Provincial Archives of Alberta, Edmonton
PAM: Provincial Archives of Manitoba, Winnipeg
PIMA files: Prairie Implement Manufacturers Association office files, Regina
PMA: Provincial Museum of Alberta, Edmonton
SABR: Saskatchewan Archives Board, Regina
SABS: Saskatchewan Archives Board, Saskatoon
UBA: University of Brandon Archives, Brandon
UMA: University of Manitoba Archives, Winnipeg
WDMS: Western Development Museum, Saskatoon

Introduction

1. For a good survey of prairie farming and its technology, see Dave Richeson, "Farming Moves into High Gear," *Horizon Canada* 1987, 2830.
2. Some analysts use four categories, each distinguished by the range of production from greatest to fewest products: full line, long line, short line, and short-short line. We adopt the more general division of full and short line.

The Prairie Region *(pages 3–17)*

1. J.H. Richards, "Retrospect and Prospect," in *Studies in Canadian Geography*, ed. P.J. Smith (Toronto: University of Toronto Press, 1972), 118.
2. Sarah Carter, *Lost Harvests: Prairie Indian Reserve Farmers and Government Policy* (Montreal: McGill-Queens University Press, 1990), 38–39.
3. W.L. Morton, "Agriculture in the Red River Colony," *Canadian Historical Review* 30 (1949): 305–21.
4. Gerald Friesen, *The Canadian Prairies: A History* (Toronto: University of Toronto Press, 1984), 179.
5. Ibid., 178–80.
6. Ibid., 184.
7. Ibid., 188–89.
8. *Census of Canada 1891,* 16; Friesen, *The Canadian Prairies,* 185.
9. W.L. Morton, *Manitoba: A History* (Toronto: University of Toronto Press, 1957), 151–201.
10. Friesen, *The Canadian Prairies*, 202.
11. Coping with dry land conditions through irrigation took place in southern Alberta,

especially the large corporate irrigation projects in the Lethbridge area that began in 1900. Irrigated farming was never typical of the region, however, although by 1980 there were some irrigated lands in Saskatchewan and Manitoba as well.

12. Friesen, *The Canadian Prairies*, 328.
13. Ibid., 301, 329.
14. Ibid., 328–29; John Herd Thompson, *The Harvests of War: The Prairie West 1914–1918* (Toronto: McClelland and Stewart, 1978).
15. Morton, *Manitoba: A History*, 208, 257, 297; *Farm and Ranch Review*, August 1932.
16. Friesen, *The Canadian Prairies*, 328.
17. Richards, "Retrospect and Prospect," 127.
18. Friesen, *The Canadian Prairies*, 329.
19. Robert E. Ankli, "Farm Income on the Great Plains and Canadian Prairies 1920–1940," *Agricultural History* 51 (1979): 92–103.
20. Friesen, *The Canadian Prairies*, 388.
21. *The Nor'-West Farmer*, December 20, 1923.
22. Vernon Fowke, "Economic Effects of the War on the Prairie Economy," *Canadian Journal of Economics and Political Science* XI (1945): 386.
23. Bruce Proudfoot, "Agriculture," in *Studies in Canadian Geography*, ed. P.J. Smith (Toronto: University of Toronto Press, 1972), 61.
24. Rex E. Grose, "Manufacturing in Manitoba: Growth and Trends," *Canadian Geographical Journal* (March 1953): 85; Proudfoot, "Agriculture," 61–62. Sugar beets were grown in southern Manitoba by the early 1930s and were sent to Minnesota for processing. A sugar beet factory was built in Winnipeg in 1940; this was the true beginning of the industry in Manitoba.
25. Proudfoot, "Agriculture," 63–64.
26. Lyle Dick, *Farmers "Making Good" The Development of Abernathy District, Saskatchewan 1880–1920* (Ottawa: Parks Canada, 1989), 52–68.
27. Friesen, *The Canadian Prairies*, 319.
28. Ibid., 309.
29. Ibid., 384.
30. *The Case For Alberta. Part I - Alberta's Problems and Dominion-Provincial Relations* (Edmonton: King's Printer, 1938), 120–21.
31. Baldur H. Kristjanson, "The Capital Structure of Canadian Agriculture," *Economic Analyst* 27 (1957): 57–61 (Part I) and 101–105 (Part II); Proudfoot, "Agriculture," 57–58.
32. A.W. Mahaffy, *The Machine Process in Agriculture With Special Reference to Western Canada* (Saskatoon: University of Saskatchewan, Department of Economics, 1923), 37–40.
33. Census of the Prairie Provinces 1946, Table 1, p. 4.
34. Proudfoot, "Agriculture," 55–56.
35. *The Commercial*, July 24, 1883, and March 23, 1886.
36. *Annual Report, Department of Agriculture of the Province of Saskatchewan, 1907*, 135–37.
37. *The Grain Growers' Guide*, February 11, 1920.
38. *Canadian Annual Review of Public Affairs 1920, 145; Canadian Farm Implements, April 1912.*
39. *Report on the Manufacturing Statistics of the Prairie Provinces, 1927* (Ottawa: Canada Department of Trade and Commerce, DBS Census of Industry, 1929).
40. *Blairmore Enterprise*, November 17, 1927 (syndicated article).
41. *The Western Producer*, December 27, 1928.
42. Friesen, *The Canadian Prairies*, 383.
43. Brenton M. Barr, "Reorganization of the Economy Since 1945," in *Studies in Canadian*

Geography, ed. P.J. Smith (Toronto: University of Toronto Press, 1972), 75–77; "Manufacturing Prospects From a Regional Perspective," Department of Regional Economic Expansion Canada, 1978, 23–28, Papers of the Alberta Department of Economic Development, 90.91/4, PAA.

44. Barr, "Reorganization of the Economy," 70–73.
45. *The Commercial*, July 24, 1883.
46. *The Grain Growers' Guide*, May 27, 1914.
47. For example, see *Canadian Farm Implements*, April 1939.
48. "An Industrial Development Strategy for Saskatchewan 1976," p. 5, Minister's Papers, Saskatchewan Department of Agriculture, R1138, III–14C, SABR.
49. Terence J. Fay, *Farm Implement Firms as Symptomatic of the Rise of Regional Grain Cities: Winnipeg and Minneapolis, 1876–1926* (Winnipeg: Institute of Urban Studies, University of Winnipeg, 1980), 24.
50. Friesen, *The Canadian Prairies*, 191.
51. S.N. Kulshrestha and D. Devine, "The Crowsnest Rate and Prairie Development. 1. A Historical Review" (Saskatoon: University of Saskatchewan, Agricultural Scene, Publication No. 298A, 1978), 3–4.
52. *Canadian Farm Implements*, April 1939; *Regina Leader-Post*, March 6, 1939; *Country Guide*, August 1950.
53. Thomas T. Schweitzer, *Old Myths and New Choices: Railway Freight Rates and Western Economic Development* (Ottawa: Economic Council of Canada, Discussion Paper No. 269, 1984), ii, 19–31.
54. Kenneth H. Norrie, "Some Comments on Prairie Economic Alienation," *Canadian Public Policy* 2 (1976): 211–24.

The Farm Implement Industry *(pages 18–30)*

1. *Final Report on Farm Implement Inquiry* (Select Special Committee, Legislature of the Province of Saskatchewan. Regina: Queen's Printer, 1952), Table II.
2. J.D. Woods and Gordon Ltd., *The Canadian Agricultural Machinery Industry* (Ottawa: Royal Commission on Canada's Economic Prospects, 1956), 19.
3. Ibid., 16.
4. A chart showing these developments in North America is in W.J. Phillips, *The Agricultural Implement Industry. A Study in Competition* (Toronto: University of Toronto Press, 1956), 12.
5. *United States Federal Trade Commission Inquiry, 1936*, p. 7, Papers of the Farm Implement Inquiry, R283, SP124, SABR.
6. *Canadian Farm Implements*, August 1938.
7. Woods and Gordon Ltd., *The Canadian Agricultural Machinery Industry*, 3.
8. *Canadian Farm Implements*, August 1909.
9. Woods and Gordon Ltd., *The Canadian Agricultural Machinery Industry*, 4.
10. Ibid., 2.
11. Ibid., 3–5.
12. *Canadian Farm Implements*, August 1936.
13. *The Commercial*, October 23, 1893.
14. *The Nor'-West Farmer*, December 20, 1902.
15. Woods and Gordon Ltd., *The Canadian Agricultural Machinery Industry*, 6–7.
16. Ibid., 13–14.
17. Ibid., 35. On this practice before 1944, see W.J. Broehl, *John Deere's Company: A History of Deere and Company and Its Times* (Doubleday and Co. Inc., 1984), 511–12.
18. Woods and Gordon Ltd., *The Canadian Agricultural Machinery Industry*, 35.
19. Ibid., 10–13.
20. Ibid., 16.

21. *Financial Post*, December 25, 1965.
22. Phillips to MacDonald, August 23, 1968, Papers of the Royal Commission on Farm Machinery [Barber Commission], RG33/91, Vol. 4 file 1–7–2, NAC.
23. "Industry Profile: Agricultural Machinery" (Ottawa: Industry, Science and Technology Canada, 1988). In 1971 Canadian production of farm equipment was almost 8 percent of total North American production. Sales in Canada represented 12 percent of that continental market (*House of Commons Debates*, December 14, 1971, 10478). This Canadian share of North American production varied between 4.8 percent and 9.4 percent between 1950 and 1974 (A.N. Brook and P. Kampouris, *Statistics Relating to Farm Machinery in Canada 1950–1976* [Ottawa: Agriculture Canada, 1977], 8).
24. Woods and Gordon Ltd., *The Canadian Agricultural Machinery Industry*, 24–26. Between 1945 and 1953, production of farm equipment in the United States ranged between 10 and 19 times that of Canada.
25. Quoted in *Final Report on Farm Implement Industry* (1952 Saskatchewan), 33; A. Stewart, "The Report of the Special Committee on Farm Implement Prices 1937," *Canadian Journal of Economics and Political Science* 4 (1937):92.
26. *Country Guide*, June 1938; *Canadian Farm Implements*, March and April 1939.
27. *Canadian Farm Implements*, April 1939; *Final Report on Farm Implement Industry* (1952, Saskatchewan), 35.
28. Report to CCIL Directors, 1956, CCIL Papers, P846, PAM.
29. "The Farm Machinery Industry in Manitoba" (1972), p. 13, Papers of the Manitoba Department of Industry and Commerce, ITT 0004, Box 20, PAM; Report to CCIL Directors, 1959, P846, PAM.
30. Memorandum, June 8, 1967, RG33/91, Vol 4 file 1–7–2, NAC.
31. *The Nor'-West Farmer*, September 1891.
32. D. Martinusen and B. Barry, *Revenues, Costs and Profits in the Farm Machinery Industry* (Ottawa: Royal Commission on Farm Machinery, 1970), 109.
33. *Submission to the Royal Commission on Farm Machinery by the Canadian Co-operative Implements Ltd., Ottawa, December 13, 1967*, CCIL Papers, P847, PAM; *Regina Leader-Post*, May 23, 1961.
34. Stewart, "Report of the Special Committee on Farm Implement Prices," 94.
35. Macdonald to Brown, September 20, 1944, Douglas Papers, 33.2, file 223, SABR.
36. Phillips, *The Agricultural Implement Industry*, 12.
37. *Canadian Farm Implements*, January, August, and October 1937.
38. *Regina Leader-Post*, February 13, 1970.
39. *The Western Producer*, December 3, 1981. For the Winnipeg exhibition, see *The Nor'-West Farmer*, November 1891, and for Brandon, see *The Nor'-West Farmer*, August 5, 1915.
40. *Canadian Farm Implements*, March 1939; *Regina Leader-Post*, March 11, 1952.
41. *Canadian Farm Implements*, March 1939.
42. *Canadian Farm Implements*, January 1937.
43. *The Nor'-West Farmer*, January 20, 1905; *Canadian Farm Implements*, March and October 1908, May 1927, March 1939, and October 1945; *Regina Leader-Post*, March 11, 1952.
44. Report of CCIL Directors, 1956, P846, PAM.
45. *Regina Leader-Post*, May 18, 1971.
46. *Canadian Farm Implements*, September 1936 and June 1939.
47. *Canadian Farm Implements*, May 1937 and April 1939.
48. *Country Guide*, June 1953; *Financial Post*, February 1965.
49. *Final Report on Farm Implement Industry* (1952, Saskatchewan), 15–16.
50. Ibid.; *Canadian Farm Implements*, May 1961; Martinusen and Barry, *Revenues, Costs and Profits in the Farm Machinery Industry*, 19–20.

Farmers and the Full Line Industry *(pages 31–42)*

(pages 31–42)

1. *The Nor'-West Farmer*, June 21, 1915.
2. *Farm and Ranch Review*, March 5, 1914.
3. *Cardston Globe*, February 27, 1913.
4. *The Grain Growers' Guide*, April 9, 1913.
5. *The Nor'-West Farmer*, October 20, 1919.
6. *Canadian Farm Implements*, June 1958.
7. *The Western Producer*, September 3, 1959; Interview with John Buhler, September 30, 1990; W.S. Good and M.D. Beckman, "A Prairie Distribution System in Transition: The Case of Farm Machinery Parts in Manitoba," *Prairie Forum* 2 (1977): 43–45.
8. Typescript for "Rural Councillor," 1973, R1138, II–3, SABR.
9. Killbery Industries, November 20, 1972, Canada Department of Industry, Trade and Commerce Papers, RG20, V 2040 file V80001–3710M1, NAC; Minutes, Manitoba Farm Machinery Board, June 24, 1974, and Minutes, Meeting with Versatile, April 23, 1973, Manitoba Department of Agriculture Papers, AG0302, Box 1, PAM; *Canadian Farm Equipment Dealer*, March 1974; *Canadian Farm Implements*, January 1965.
10. Kirchner Machine Ltd., RG20, V 2040 file 8001–3710M1, NAC.
11. *Submission to the Royal Commission by CCIL 1967*, P847, PAM.
12. *Fifty Years of Progress on Dominion Experimental Farms, 1886–1936* (Ottawa: King's Printer, 1939), 77, 138–39; *Annual Report, Dominion Experimental Station, Swift Current, 1931*, 83; Correspondence, 1937–38, C.S. Noble Papers, M7198, files 19 and 21, GAI. Testing at Swift Current was temporarily suspended in 1931 because of the Depression.
13. See, for example, Minutes, Agricultural Engineering Committee, January 25, 1936, November 18 and 19, 1949, Alberta Department of Agriculture Papers, 73.307/374, PAA; Session No. 5, February 5, 1951, Agriculture Canada, Central Registry Files, 1918–53, RG17, Vol. 2893 Reel 1841, NAC.
14. For example, *Country Guide*, July 1934.
15. *Canadian Farm Implements*, November 1962.
16. *Annual Report, AMA, 1960/61*, 22–25.
17. J.T. Kyle, "Farm Machinery Testing by Public Agencies," Address to the ASAE 1962, Papers of the Saskatchewan Agricultural Machinery Administration, R372, fII–6, SABR.
18. Interview with W.B. Reed, May 19, 1990.
19. Brown to Nollet, March 11, 1959, Douglas Papers, 33.1 file 60, SABR.
20. Horner to Gardner, April 30, 1965, R372, fII–6, SABR.
21. *Regina Leader-Post*, December 7, 1965, and February 21, 1966; Resume of the Research and Development Section, Department of Agricultural Engineering (n.d.), R1138, IV–22B, SABR.
22. *Winnipeg Tribune*, March 30, 1967.
23. *Regina Leader-Post*, September 10, 1971.
24. Comments on Barber Commission, May 19, 1971, and Messer to Olson, June 27, 1972, R1138, II–11A, SABR; *Regina Leader-Post*, May 19, 1971; *Annual Report, PAMI, 1975*, not paginated.
25. *Annual Report, PAMI, 1988/89* (not paginated); *Canadian Farm Equipment Dealer*, March 1975; *Regina Leader-Post*, March 3, 1977.
26. *Regina Leader-Post*, June 24, 1978.
27. *The Nor'-West Farmer*, December 20, 1902, and February 5, 1920; *Farm and Ranch Review*, March 5, 1914; *Canadian Farm Implements*, April 1939; *The Country Guide*, October 1953.
28. "Farm Machinery Syndicate Loans" (1964 pamphlet), Alberta Wheat Pool Papers, M2369, file 139, GAI; B.H. Sonntag, "Farm Machinery Syndicates in the Prairie Provinces," *Canadian Farm Economics* (August 1966) 14.

29. "Report of Interprovincial Meeting to Consider the Handling of Farm Implements," 73.307/379, PAA; *Mechanization and Farm Costs* (Royal Commission on Agriculture and Rural Life, Regina: Queen's Printer, 1955), 16.
30. See, for example, Davis to Graham, March 10, 1939, R283, SP No. 124, SABR; *Final Report on Farm Implement Industry* (1952, Saskatchewan), 15–16; *Regina Leader-Post*, May 18, 1971.
31. For analysis of this issue, see K. Norrie, "Agricultural Implement Tariffs, the National Policy and Income Redistribution in the Wheat Economy," *Canadian Journal of Economics* 7 (1974): 449–62.
32. *Farm and Ranch Review*, March 5, 1914.
33. *The Grain Growers' Guide*, September 20, 1911.
34. *The Nor'-West Farmer*, June 21, 1915.
35. Minute Book, Board of Directors, December 10, 1918, and December 10, 1920, Minutes, Annual General Meeting 1920, p. 7, and Minutes, Annual General Meeting 1921, United Grain Growers' Papers, A90–35/20, UMA; *United News*, February 1920, 2–3.
36. "Memorandum on Farm Machines, Their Manufacture, Distribution and Prices. Based on the Experience of the CCIL, October 24–25, 1957," 2–4, P847, PAM.
37. *Final Report on Farm Implement Industry* (1952, Saskatchewan), 19. On the early formation of CCIL, see also *Co-operative Leadership: Harry Fowler's Memoirs*, Introduction by T. Phalen (Saskatoon: Co-operative College of Canada, 1977), 137–51; and Ian MacPherson, "'Better Tractors for Less Money': The Establishment of Canadian Co-operative Implements Ltd.," *Manitoba History* 13 (1987): 2–11.
38. Minutes of Directors' Meeting, Interim Report (ca. July 1940), CCIL Papers, P2509, PAM; *Final Report on Farm Implement Industry* (1952, Saskatchewan), 23.
39. Report to CCIL Directors, February 24–25, 1949, P846, PAM.
40. "Submission to the Royal Commission on Farm Machinery by CCIL." Cockshutt purchased National's assets after National's bankruptcy in 1952.
41. *Final Report on Farm Implement Industry* (1952, Saskatchewan), 25–26, 36; "Memorandum on Farm Machines, Their Manufacture, Distribution and Prices. Based on the Experience of the CCIL," 24.
42. Directors' Report, CCIL, February 24–25, 1949, P846, PAM.
43. *Final Report on Farm Implement Inquiry* (1952, Saskatchewan), 23.
44. Arnason to Barker, June 10, 1954, Papers of the Royal Commission on Agriculture and Rural Life, R236 Report No. 2, SABR.
45. Phillips, *The Agricultural Implement Industry*, 141; *Canadian Farm Implements*, March 1939.
46. *The Western Producer*, May 12, 1988.
47. *Final Report on Farm Implement Industry* (1952, Saskatchewan), 25.
48. Ibid., 24–25.
49. Report to CCIL Directors, 1955, P846, PAM.
50. On CCIL's development to the late 1960s, see Rubin Simkin, *The Prairie Farm Machinery Co-operative: The "CCIL"* (Ottawa: Queen's Printer, 1970).
51. *Final Report on Farm Implement Industry* (1952, Saskatchewan), 20–21.
52. *Canadian Farm Equipment Dealer*, March 1979.
53. President's Report to CCIL Directors, November 28–29, 1968, P2509, PAM.
54. Hearings at Calgary, March 16, 1967, p. 469, RG33/91, Vol. 16 f. Ex–62–64, NAC; Testimony by National Farmers' Union, January 16, 17, 1968, p. 55, and file 1–7–2, RG33/91, NAC; *Annual Report, CCIL 1968*, 8.
55. Manitoba Farmers' Union Brief to the Royal Commission on Farm Machinery, March 7, 1967, 15, RG33/91, Vol. 13 Ex–4, NAC.
56. *Annual Report, CCIL 1967*, 7; President's Report to CCIL Directors, November 28–29, 1968, P2509, PAM.

57. Report, Directors Meeting, June 10–11, 1968, P2509, PAM.
58. Report to Annual CCIL Meeting, 1973, CCIL Papers, P2510, PAM.
59. See, for example, *Annual Report, CCIL, 1969*, 4.
60. *Regina Leader-Post*, October 4, 1980.
61. *Winnipeg Tribune*, April 12, 1978.
62. *The Globe and Mail*, March 9, 1991; Interview with Walter Dressel, September 29, 1990.

An Early Prairie Industry *(pages 43–55)*

1. *Agricultural Implement Industry 1918* (Ottawa: Department of Trade and Commerce, DBS Census of Industry, 1920); *Census of Canada 1901*, Vol. 3, Table XXIV, pp. 362–63 and *Census of Canada 1911*, Vol. 3, Table V, pp. 120–32.
2. The data was compiled largely from these three publications, but any additional references found during research were also added. Use was also made of the Canadian Agricultural Manufacturers Index at the National Museum of Science and Technology, Ottawa. On this index, see Tom Brown, "Canadian Agricultural Equipment Manufacturers Advertisement Index, 1847–1942," *Material History Bulletin*, Spring 1987, 29–33.
3. *Census of the Northwest Provinces, 1906*, Table 1, p. 17; *Report on the Manufacturing Statistics of the Prairie Provinces 1927*, (Ottawa: Canada Department of Trade and Commerce, DBS, Census of Industry, 1929), Table 4, pp. 7–13.
4. *The Grain Growers' Guide*, February 11, 1920.
5. "Observations on Machine Shop Activity in Saskatchewan," April 1957, R372, II–47, SABR.
6. *The Commercial*, October 3 and October 24, 1882.
7. *Canadian Farm Implements,* February 1946. On Cater, see W. Leland Clark, "Harry Cater: The Personification of the Successful Municipal Politician?" *Historical and Scientific Society of Manitoba, Transactions*, Series III (1977–79): 163–76.
8. "The Farm Implements and Machinery Industry in Canada 1946" (DBS, Census of Industry, 1946), Table 2.
9. *Census of Canada 1951,* Table 13 ("Farm Machinery 1921–1951" for each province). Automobiles were the second most common machines after binders.
10. *The Commercial*, October 3, 1882.
11. Noble to Hutchings, February 12, 1936, M7198 file 41, GAI.
12. *Annual Report, Dominion Experimental Station, Swift Current, 1934*, 134–35.
13. Minutes, Agricultural Engineering Committee, January 25, 1936, 73.307/374, PAA.
14. *Made in Saskatchewan: A Story of Invention. Index 1905–1980* (Saskatoon: Western Development Museum, 1980), v. Working lists of patents granted to Albertans for agricultural equipment have been compiled by the Reynolds Alberta Museum, Wetaskiwin.
15. For rates, see Fetherstonaugh and Co., December 8, 1936, M7198 file 24, GAI. The fees in 1936 in Canada were $125.00 and up.
16. For example, *The Nor'-West Farmer*, January 20, 1919.
17. Helmer H. Hanson, "History of Swathing and Swath Threshing," RG33/91, Vol. 21 file 11–2–1 Vol. 8, NAC.
18. Ibid.
19. Evan Hardy, "Progress in Agricultural Engineering in Western Canada," 1947 Conference Paper, Faculty Publications, University of Saskatchewan Archives.
20. *Canadian Farm Implements*, April 1939.
21. Hanson, "History of Swathing and Swath Threshing."
22. *The Commercial*, November 23, 1893.
23. *Medicine Hat Mail*, January 19, 1913.
24. *The Grain Growers' Guide*, February 11, 1920.
25. "Gilbert Harvester Stacker" (1937 pamphlet), 73.307/378, PAA.

26. Davies to Noble, May 29, 1937; Garnett to Noble, May 26, 1937, and Banting to Noble, May 22, 1937, M7198, File 8, GAI.
27. Morton, *Manitoba: A History*, 395, 455.
28. *Canadian Farm Implements*, April 1939.
29. *Independence*, August 22, 1901. See also *Lethbridge Herald*, April 2, 1910.
30. J.W. Morrison, "Agricultural Achievements of the Prairie Experimental Farms 1886–1986," *Prairie Forum* 11 (1986): 195–213.
31. Morton, *Manitoba: A History*, 395.
32. *Regina Leader-Post*, February 16, 1939.
33. *The Nor'-West Farmer*, January 6, 1919.
34. Circular Letter, ca. 1920, Canadian Farm Implement Co. Ltd., WDMS Files.
35. For example, see *The Grain Growers' Guide*, June 2, 1915.
36. Memorandum of Association, The Seed Grain Grader Co., July 21, 1915, Saskatchewan Corporate Registry, 16, File 001988, SABR. The company was struck off the corporate registry in 1918.
37. Hearings, Prince Albert, June 19, 1914, pp. 22–25, Papers of the Saskatchewan Farm Machinery Commission 1914, R243, SABR.
38. *Canadian Farm Implements*, August 1939.
39. Irving to Noble, December 29, 1937, M7198, File 49, GAI.
40. *Prize List 1912*, p. 5, Brandon Fair Board Records, BUA.
41. *Canadian Farm Implements*, July 1908. On other exhibits by local farm equipment firms, see *The Nor'-West Farmer*, August 21, 1905.
42. *Canadian Farm Implements*, March 1938 and July 1939.
43. "Gilbert Harvester-Stacker" (1937 pamphlet), 73.307/378, PAA.
44. Ross Irwin, "The Sylvester Auto-Thresher: A Canadian First in Grain Harvesting," *Scientia Canadensis* 8 (1984): 144.
45. Minutes, UGG Annual Meeting 1921, pp. 20–21, A90–35/20, UMA.

Growth and Development . . . *(pages 56–72)*

1. *Canadian Farm Implements*, November 1958.
2. *Winnipeg Tribune*, February 22, 1980; *Regina Leader-Post*, June 2, 1979.
3. Head to Director, February 4, 1971, RG20, Vol. 2075 File U8001–3710/M1, Vol. 7, NAC.
4. Phillips, *The Agricultural Implement Industry*, 44.
5. Hayes to Sleigh, October 3, 1967, ITT 0004, Box 20, PAM. See also, hearings at Calgary, March 17, 1967, RG33/91, Vol. 7, 690, NAC.
6. Company Profile, November 10, 1988, typescript; *Small Business*, June 1987; Interview with Jim Carnago, May 23, 1990.
7. Interview with Ray Malinowski, June 1, 1990.
8. Transcript of Hearings, October 13, 1953, R236 Report No. 2, SABR.
9. *Regina Leader-Post*, October 13, 1973.
10. *The Western Producer*, December 20, 1962.
11. Interview with Paul Soubry, September 28, 1990.
12. *Winnipeg Tribune*, February 16, 1967; *Regina Leader-Post*, November 12, 1969; *Financial Post*, June 26, 1976.
13. McKee to Barber Royal Commission, September 21, 1966, RG 33/91, Vol. 4 file 1–7–2, NAC.
14. *Annual Report, Co-op Implements, 1974/75*, pp. 36–37 and 1975/76, pp. 28–29.
15. *Financial Post*, October 9, 1976.
16. Interview with Paul Soubry, September 28, 1990.
17. Hearings at Calgary, March 16, 1967, RG 33/91, Vol. 16, 497, NAC.
18. *Financial Post*, January 4, 1969.

19. "Alberta Manufacturing Opportunities, Agricultural Implements," ca. 1965, RG20, V2040 file V8001–M3710/M1 Vol. 12, NAC; Interview with Francis Doepker, May 22, 1990.
20. Booz Allen and Hamilton Canada Ltd., "Agricultural Implement Manufacturing Industry Study," 1972, p. 4, ITT 0004, Box 19, PAM.
21. "The Morris Story," Prepared by Western Construction and Industry Magazine in Conjunction with the Morris Rod Weeder Company 50th Anniversary, March 15, 1979, p. 9.
22. *Canadian Farm Equipment Dealer*, March/April 1976.
23. *Canadian Farm Implements*, November 1960.
24. Report to CCIL Directors, 1968, P846, PAM.
25. Pig iron (produced by smelting ore) was too expensive to use at Selkirk because of the high cost of importing it and suitable coal and coke. Prairie consumption of sheet iron and heavy structural steel was too low to make such production with pig iron feasible, nor was there sufficient scrap for such production. In 1974, the mill still did not produce sheet steel. Arthur D. Little Inc., "Feasibility of New Iron and Steel Capacity in the Province of Manitoba," Report to the Manitoba Department of Industry and Commerce, 1954, pp. 1–2 (Copy in National Library, Ottawa); "Task Force on the Agricultural Equipment Sector: Information on Manufacturing Currently Available," 1975, p. 6, ITT 0004, Box 20, PAM.
26. John Archer, *Saskatchewan: A History* (Saskatoon: Western Producer Prairie Books, 1980), 311.
27. *Regina Leader-Post*, October 13, 1973.
28. *Regina Leader-Post*, September 16, 1974.
29. *Canadian Farm Implements*, August 1954.
30. S. Andrew and A.R. Brown, "Farm Electric Plants in Alberta," (Edmonton: Alberta Power Commission, 1944 pamphlet); D. Gayton, "Wind Power on the Prairies," 1987, Clipping file "Prairie History," Regina Public Library.
31. Interview with Clem Roles, May 20, 1990. Smith sold his interest in the company to Roles in 1949.
32. Les Edwards, *An Idea and Fourteen Dollars* (Saskatoon: Smith-Roles Ltd, (ca. 1980), not paginated; *The Record Stockman*, May 27, 1982; Clem Roles, *Better Arc Welding* (Saskatoon: Smith-Roles Ltd, 1964).
33. "Observations on Machine Shop Activities in Saskatchewan," 1957, R372, II–47, SABR.
34. *Agriculture and Industrial Progress in Canada*, November 1947, 173.
35. "Alberta Manufacturing Opportunities, Agricultural Implements," ca. 1965, RG20, V2040 file V8001–M3710/M1 Vol. 12, NAC.
36. Hearings, March 16, 1967, RG33/91, Vol. 16, pp. 112–13, 118, NAC; *Winnipeg Tribune*, November 16, 1979.
37. Interview with Ray Malinowski, June 1, 1990.
38. *Winnipeg Tribune*, December 14, 1967.
39. *Regina Leader-Post*, September 16, 1974.
40. P.S. Ross and Partners, "Export Opportunities in the Upper Midwest United States for Manitoba Manufactured Agricultural Equipment," Report for Manitoba Department of Industry and Commerce, 1968 (copy in Manitoba Legislature Library), 7.
41. "Research and Economic Development 1953," R33.2, file 75, SABR.
42. Minutes, Meeting of Industry and Resources Development Committee, May 16, August 14, August 12, 1952, and August 7, 1953, R33.1, file 82, SABR.
43. Archer, *Saskatchewan: A History*, 311.
44. "Province of Manitoba Brief to Royal Commission on Farm Machinery, March 6, 1967," RG33/91, Vol. 13 Ex–2, NAC.
45. *Winnipeg Tribune*, August 10, 1978.
46. Martinusen and Barry, *Revenues, Costs and Profits in the Farm Machinery Industry*, 89.

47. Robin Nodwell Manufacturing Ltd. Presentation to the Royal Commission on Farm Machinery, 1967 (copy at University of Calgary Library). For the situation in Manitoba, see P.S. Ross Partners, "Export Opportunities in the Upper Midwest United States." For Alberta, see Erikson to Vanterpool, November 17, 1976 (attachments), Alberta Department of Economic Development Papers, 83.372/8, PAA.
48. *The Western Producer*, August 4, 1983.
49. Interview with Ray Malinowski, June 1, 1990; Interview with Jim Carnago, May 23, 1990.
50. "The Markets and Demand For/Supply of Equipment, 1977," Alberta Department of Economic Development Papers, 86.521/9, PAA.
51. *Winnipeg Tribune*, February 16, 1967.
52. *Canadian Farm Equipment Dealer*, April 1971.
53. A.N. Seifried, "The Westward Shift of Manufacturing on the Prairies," *Prairie Forum*, 1 (1986):99.
54. *Financial Post*, December 6, 1980.
55. "Province of Manitoba Brief to Royal Commission on Farm Machinery March 6, 1967," RG33/91, Vol. 13 Ex–2, NAC.
56. Hearings March 16, 1967, RG33/19, Vol. 16, p. 113, NAC; Robinson to Barber, June 9, 1967, RG33/91, Vol. 5 file 1–7–2, NAC.
57. Kirchner Machine Ltd., RG20, V2040 V8001–3710M1, NAC.
58. For example, *Regina Leader-Post*, August 12, 1963; *Winnipeg Tribune*, January 18, 1972.
59. "Activities . . . Canadian Government . . . Shortline Agricultural Equipment," ca. 1970, RG20, V2075 f.V8001–3710/M1, NAC. The value of these export programs remains unclear. One study of Alberta's programs notes that there is little empirical evidence of their effectiveness. Alberta businesses tended to use federal programs over provincial ones for assistance in reducing obstacles to exports (A.O. Ajao and B.M. Barr, "Entrepreneurial Assessment of Government Programmes to Encourage Alberta's Foreign Trade," *Prairie Forum*, 2 [1977]: 154–63.
60. "Versatile History," typescript ca. 1981, courtesy Peter Pakosh.
61. *Financial Post*, November 24, 1979.
62. *The Western Producer*, September 13, 1979.
63. *Financial Post*, November 24, 1979.
64. Report on the Agricultural Machinery Administration, 1964, R372, fII–6, SABR.
65. For example, Anderson Equipment, November 22, 1972, RG20, V2040 file V8001–3710M1, NAC.
66. "The Morris Story," 7.
67. Williams to Purnell, October 16, 1972 (attachment), Alberta Department of Agriculture Papers, 86.423/14, PAA.
68. For example, Douglas to Hayes, December 30, 1969, ITT 0004, Box 24, PAM; Payne-Ross Ltd., "Committee on Manitoba's Economic Future. Study of Agricultural Supplies and Equipment Industry (1962)," Manitoba Department of Industry, Trade and Tourism Papers, RG6, D.1, PAM.
69. *Canadian Farm Equipment Dealer*, July 1968.
70. P.S. Ross and Partners, "Export Markets in the Upper Midwest," 32.
71. Hearings March 16, 1967, RG33/91, Vol. 16, p. 113, NAC; Robinson to Barber, June 9, 1967, RG33/91, Vol. 5 file 1–7–2, NAC.
72. Interview with Ross Giles, May 28, 1990.
73. *Foreign Trade*, March 18, 1967; *Canadian Farm Equipment Dealer*, April 1971 and April 1972.
74. "Farm Machinery Show Survey (1969)," ITT 0004, Box 16, PAM.
75. Hearings at Regina, March 28, 1967, RG33/91, Vol. 16, NAC; "Morris Soil Conservation

School and Product Demonstration" (pamphlet), 1971, File VIII–2, WDMS; *Canadian Farm Equipment Dealer*, July 1972.
76. *Regina Leader-Post*, June 22 and June 26, 1978.
77. Martinusen and Barry, *Revenues, Costs and Profits in the Farm Machinery Industry*, 97, Table 7–2, 113, 127–28.
78. "Versatile History" (1981); *Winnipeg Free Press*, January 3, 1990; Interview with Paul Soubry, September 28, 1990.
79. Report of ASAE Meeting, December 6–8, 1943, RG17, Vol. 3216 Reel 155–2, NAC.

Organizing the Industry . . . *(pages 73–85)*

1. Ivan Thue, *The PIMA Story* (Regina: PIMA, 1990), 2.
2. *Minutes*, Manitoba Wholesale Implements Association, various entries January 17, 1935–1984, Minute Book (Manitoba Wholesale Implements Association Offices, Winnipeg).
3. Torian to Scharf, December 9, 1969, PIMA files.
4. Minutes, Meeting of Saskatchewan Farm Machinery Manufacturers, December 5, 1969; Schnell to Scharf, December 12, 1969, PIMA files.
5. CFIEI brochure, 1972, PIMA files.
6. "Province of Manitoba Brief to Royal Commission on Farm Machinery, March 6, 1967," RG33/91, vol. 13, file EX–2, p.17, NAC; "Questions and replies on the Province of Manitoba Brief," RG33/91, Vol. 16, p. 54, NAC. See also *The Western Producer*, September 7, 1972.
7. Thue, *PIMA Story*, 2–7; PIMA News Release, February 24, 1970, PIMA files.
8. Report on Meeting of February 20 and 26, 1970, PIMA files.
9. Merle Boesch notes, 1981, PIMA files; Thue, *PIMA Story*, 7, 12, 13, 28.
10. Thue, *PIMA Story*, 44–45.
11. Membership Lists, 1970–72 and 1974, PIMA files; Thue, *PIMA Story*, 18, 44; *PIMA Pulse*, March 1990.
12. *Regina Leader-Post*, March 22, 1980.
13. Thue, *PIMA Story*, 27, 38.
14. Interview, Adeleine Morris, May 30, 1990; *Regina Leader-Post*, February 27, 1978.
15. "Growth and Survival in the 70's" (pamphlet), 1971, PIMA files.
16. Thue, *PIMA Story*, 41.
17. Ibid., 18.
18. *Factory Action*, April 1973.
19. *Farm Light and Power*, June 1985; *Regina Leader-Post*, June 22, 23, and 26, 1978, June 2, 1979; Thue, *PIMA Story*, 22–24.
20. Thue, *PIMA Story*, 22–24, 44–45.
21. Ibid., 11–15, 31.
22. *PIMA Pulse*, December 1990; Thue, *PIMA Story*, 30–39; News Release, October 8, 1991, PIMA files.
23. Thue, *PIMA Story*, 42–44; *PIMA Pulse*, April, June, and November 1990 ("Innovations Bulletin").
24. Symes to Rousell, March 12, 1973, PIMA files.
25. Roles to Wilson, January 21, 1972, and Wilson to Roles, August 23, 1972, PIMA files.
26. Thue, *PIMA Story*, 13, 24; Minutes, Annual Meeting, 1974, PIMA files.
27. *Factory Action*, April 1973; Thue, *PIMA Story*, 40–45; *PIMA Pulse*, November 1991.
28. *Canadian Farm Equipment Dealer*, June 1973.
29. Minutes, PIMA Annual Convention, 1975, PIMA files.
30. Thue to PIMA regular members (circular letter), October 20, 1977, PIMA files.
31. Thue, *PIMA Story*, 22.
32. *The Western Producer*, January 22, 1981.

33. Resolutions, PIMA Annual Convention, 1973, PIMA files.
34. Rousell to Leitch, March 27, 1973, PIMA files.
35. Thue, *PIMA Story*, 8; PIMA to Schreyer, December 13, 1972, PIMA files.
36. *PIMA Pulse*, October 1990; Thue, *PIMA Story*, 22, 28–29, 37, 42.
37. E.N. Tyrchniewicz, *An Exploratory Analysis of Transport Problems Relating to the Supply of Farm Machinery and Repair Parts to Prairie Farmers* (Ottawa: Canadian Transport Commission, 1973), 19–20, 28; Thue, *PIMA Story*, 18.
38. *Regina Leader-Post*, February 27, 1978.
39. Minutes, Tri-Provincial Meetings Farm Machinery Board, November 23, 1977, AG0319, G-3-3-7, PAM.
40. Thue, *PIMA Story*, 30, 37.
41. PIMA to regular members (circular letter), May 24, 1977, PIMA files.
42. *Regina Leader-Post*, October 25, 1976; Thue, *PIMA Story*, 29.
43. *Winnipeg Free Press*, June 8, 1977.
44. *The Western Producer*, September 13, 1979.
45. Thue, *PIMA Story*, 39–41.
46. Ibid., 44.
47. *PIMA Pulse*, April 1990.
48. Minutes, Meeting of the Farm Machinery Boards of Alberta, Saskatchewan and Manitoba, August 21, 1974, PIMA files.
49. *PIMA Pulse*, November and December, 1990.

Sources of Power for Prairie Agriculture *(pages 86–106)*

1. R. Bruce Shepard, "Tractors and Combines in the Second Stage of Agricultural Mechanization on the Canadian Plains," *Prairie Forum* 11 (1986): 253–54. The distinction used here between machines and tools defines a machine, and hence mechanization, as a "combination of interrelated parts for using or applying energy to do work; especially, a mechanical unit with or without a driving motor." A tool is a "simple mechanism, or implement . . . used chiefly in manual work." An implement is therefore less specific than either a tool or machine and is "anything designed to effect a purpose."
2. David Spector, *Agriculture on the Prairies* (Ottawa: Parks Canada, 1983), 143.
3. Penny Ham, *Manitoba Agricultural Museum* (Austin, Manitoba: Manitoba Agricultural Museum, 1979), 179; Morton, *Manitoba: A History*, 207, 255.
4. *Country Guide*, May 1948.
5. Morton, *Manitoba: A History*, 393, 452.
6. *Country Guide*, May 1955.
7. Quoted in Friesen, *The Canadian Prairies*, 306–7.
8. *Country Guide*, May 1955.
9. Spector, *Agriculture on the Prairies*, 144; R. Bruce Shepard, "The Mechanized Agricultural Frontier of the Canadian Plains," *Material History Bulletin* (Spring 1979): 6.
10. *The Nor'-West Farmer*, July 5 and 20, 1932.
11. Spector, *Agriculture on the Prairies*, 144–53; O.L. Symes, "Agricultural Technology and Changing Life on the Prairies," *Seminar Proceedings, Development of Agriculture on the Prairies* (Regina: University of Regina, Canadian Plains Area Studies, 1975), 35.
12. Spector, *Agriculture on the Prairies*, 153.
13. *The Nor'-West Farmer*, July 5 and 20, 1932.
14. Spector, *Agriculture on the Prairies*, 153; O.L. Symes, "Evolution of Tillage Practices for Sustained Crop Production in the Palliser Triangle," *Transactions of the Royal Society of Canada*, Series V/Vol. 1 (1986): 99.
15. Spector, *Agriculture on the Prairies*, 152–55.
16. *The Nor'-West Farmer*, October 5, 1915.
17. Spector, *Agriculture on the Prairies*, 152–55.

18. *The Nor'-West Farmer*, July 5 and 20, 1932.
19. Ibid., February 5, 1919, July 5 and July 20, 1932; Shepard, "The Mechanized Agricultural Frontier," 17.
20. Shepard, "The Mechanized Agricultural Frontier," 17; *The Nor'-West Farmer*, July 5 and 20, 1932.
21. *The Nor'-West Farmer*, February 5, 1919; *Lethbridge Daily Herald*, March 11, 1918.
22. *Lethbridge Daily Herald*, November 29, 1913.
23. *The Nor'-West Farmer*, August 5, 1921.
24. Ibid., December 6, 1920.
25. Ibid., August 5, 1921.
26. "Recommendations of the Farm Machinery Committee (Saskatchewan) and Dominion Agronomists," Agricultural Tools and Equipment file, PMA; "The Development of Prairie Farming," n.d., Table IV, M2369, file 163, GAI.
27. For the most part, we assume that it is engine power that is being referred to when manufacturers use the term "horse power" without qualifying it.
28. *Country Guide*, May 1948; Allen to Updike, March 16, 1928, Hamilton Papers, M13 file 14, SABR.
29. Robert E. Ankli, H. Dan Helsberg, and John Herd Thompson, "The Adoption of the Gasoline Tractor in Western Canada," in *Canadian Papers in Rural History II*, ed. D. Akenson (Gananoque, Ontario: Langdale Press, 1980), 35. See also Stanley Gordon, *Agricultural Tractors in Alberta Since 1925* (Wetaskiwin: Reynolds Alberta Museum, Background Paper 17, 1983), 1–40.
30. Fred Jones, *Farm Gas Engines and Tractors,* 2d ed. (New York and London: McGraw-Hill Book Co., 1938), 327.
31. *Canadian Farm Implements*, August 1936, December 1938, and June 1939; *Country Guide*, March 1937.
32. *Canadian Farm Implements*, August 1934.
33. O.L. Symes, "Chronology of Agricultural Machinery Development and Related Information," in *Seminar Proceedings, Development of Agriculture on the Prairies* (Regina: University of Regina, Canadian Plains Area Studies, 1975); Symes, "Agricultural Technology," 22.
34. Symes, "Agricultural Technology," 37.
35. See, for example, test reports in *Annual Report, Dominion Experimental Farm, Swift Current, 1941*, 221–238.
36. "The Pattern of Mechanization and Wartime Changes on Farms in the Elrose-Rosetown-Conquest Area of Central and West Central Saskatchewan 1944," Publications of the Dominion Department of Agriculture, SABR.
37. Ibid.
38. *Regina Leader-Post*, February 29, 1952; *Country Guide*, August 1948, August 1954, and May 1955; *The Western Producer*, July 11, 1952.
39. *The Western Producer*, July 11, 1952.
40. *Financial Post*, December 25, 1965.
41. Ibid., July 1, 1967.
42. Peat Marwick, Mitchell, "Feasibility Study, Marketing and Financial . . . Romanian Universal Tractor Assembly Plant, April 1972," p. II–2, Papers of the Saskatchewan Agricultural Implements Board, 662/1, SABR; *Winnipeg Tribune*, June 16, 1977; *Regina Leader-Post*, March 1, 1979.
43. *Financial Post,* December 25, 1965; *Canadian Farm Equipment Dealer,* October 1971.
44. *Regina Leader-Post*, June 19, 1980.
45. *Alberta Report*, March 2, 1987; *Manitoba Business*, August 1984.
46. *Winnipeg Free Press*, May 3, 1975.

47. *The Nor'-West Farmer*, December 20, 1902, and January 5, 1905; *Farmer's Advocate*, July 6 and December 14, 1903 (advertisements).
48. *The Nor'-West Farmer*, April 20, 1905; *Canadian Farm Implements*, July 1908; "Manitoba Engines Ltd." (catalogue), ca. 1915, Ag 12 II Group B, file 48, SABS; *The Grain Growers' Guide*, July 2, 1919.
49. Phillips to MacDonald, August 23, 1968, RG33/91, Vol. 4 f 1–7–2, NAC.
50. Peat Marwick, Mitchell, "Feasibility Study . . . April 1972," p. IV–1 and Exhibit 2.
51. *The Nor'-West Farmer*, July 5 and 20, 1932.
52. *The Grain Growers' Guide*, February 8, 1911; *Canadian Thresherman and Farmer*, February 1911.
53. *The Nor'-West Farmer*, August 5, 1915 (advertisement).
54. K. Tingley, *The Gasoline Tractor in Western Canada: An Introduction to its History in Alberta Before 1925* (Wetaskiwin: Reynolds Alberta Museum, Background Paper 18, 1984), 148–52.
55. Massey Harris began building tractors in Canada in 1919 but stopped production in 1923. The company found that without tariff protection on small tractors, Canadian production was not competitive (Woods and Gordon Ltd., *The Canadian Agricultural Machinery Industry*, 5).
56. *Farmer's Advocate*, December 10, 1919; *The Grain Growers' Guide*, February 11, 1920 (advertisements).
57. *Financial Post*, December 25, 1965. Four wheel drive tractors were not a new concept. They had been produced by Massey from 1930 to 1936, but had significant technical problems (Gordon, *Tractors in Alberta Since 1925*, 43).
58. *Financial Post*, April 9, 1966; *Winnipeg Free Press*, December 14, 1968; Interview with Peter Pakosh, October 3, 1990.
59. Hydrostatic drive transmission uses a hydraulic pump which pressurizes oil to operate a hydrostatic motor connected to the driving wheels (*Guide to Farm Practice in Saskatchewan 1972*, 11).
60. Information on Versatile's tractors is from: *Winnipeg Tribune*, January 21, 1970; *Manitoba Business*, August 1984; *Versatile News*, October 1972; "All You Wanted to Know About Bidirectional 200 Series Tractors" (pamphlet); *Canadian Farm Equipment Dealer*, April 1971; Interview with Peter Pakosh, October 3, 1990.
61. *Regina Leader-Post*, October 16, 1970; *Saskatoon Star Phoenix*, October 17, 1974, August 2, 1975, April 7, 1979, and January 21, 1984.
62. *The Nor'-West Farmer*, August 5, 1918.
63. *The Grain Growers' Guide*, May 17, 1911; *The Nor'-West Farmer*, August 5, 1918.
64. *Agricultural Equipment Catalog*, Alberta Economic Development, 1980, p. 35.
65. *Financial Post*, May 22, 1965.
66. Fibro Industries, RG20, V 2040, V8001–3710M1, NAC.
67. *Financial Post*, November 24, 1979.
68. *Canadian Farm Implements*, March 1953.
69. Ibid.
70. The company started in 1949 as Babysh and Demkiw Farm Equipment, and with a new partner in 1953 became Demkiw and Yakimetz Equipment Ltd. (Company Profile, Typescript, n.d.; Interview with Eugene Demkiw, May 14, 1990).
71. Interview with Ray Malinowski, June 1, 1990.
72. *Farmer's Advocate*, March 20 and May 20, 1903, and July 13, 1904 (advertisements).
73. Shepard, "Tractors and Combines in the Second Stage of Agricultural Mechanization," 255.
74. "Submission to the Select Special Committee on Farm Implements by W.H. Horner, on Behalf of the Department of Agriculture, 1951," R266, file 27, SABR.

Clearing and Tilling the Land *(pages 107–38)*

1. Tillage consists of primary and secondary operations. The former requires machines for major soil tillage, the latter for finishing operations.
2. *The Nor'-West Farmer*, July 5 and 20, 1932.
3. Symes, "Evolution of Tillage Practices," 98.
4. *The Nor'-West Farmer*, July 5 and 20, 1932.
5. *The Nor'-West Farmer*, January 5, 1918; *Agricultural Alberta*, April 1921.
6. For example, *The Nor'-West Farmer,* April 20, 1920.
7. Barry Rogers and D. Tracy Anderson, *The History, Development and Use of Tillage and Cultivation Implements in Alberta Agriculture, 1870–1955* (Wetaskiwin: Reynolds Alberta Museum, Background Paper 26, 1985), 58–59; *The Grain Growers' Guide*, May 1, 1912.
8. *The Nor'-West Farmer*, May 1933.
9. Elise A. Corbet, *The Manufacture of Agricultural Implements in Alberta 1870 to 1955* (Wetaskiwin: Reynolds Alberta Museum, Background Paper 20, 1984), 65, 125–29.
10. "Costs of Clearing and Breaking Bush Lands in Alberta," *Economic Annalist*, 14 (February 1944):23; M.E. Andal, "Use of Power Machinery in Bush Land Improvement in North Eastern Saskatchewan," *Economic Annalist* 17 (May 1947):32–37.
11. *Edmonton Journal*, March 1, 1977; Company Profile (typescript), 1989.
12. *The Western Producer,* August 11, 1977, and October 26, 1989.
13. For an example of such instructions, see *The Nor'-West Farmer*, August 20, 1931.
14. *Free Press Weekly*, August 7, 1963.
15. *Canadian Farm Equipment Dealer*, April 1972.
16. *Regina Leader-Post*, April 3 and October 13, 1973.
17. Company Profile (typescript), November 10, 1988; *Small Business*, June 1987.
18. *Regina Leader-Post*, August 9, 1976.
19. *Saskatoon Star Phoenix*, April 30, 1990.
20. Interview with Ross Giles, May 28, 1990.
21. Morton, *Manitoba: A History*, 223–24.
22. Alex Johnston, "History of Agriculture in the Prairie Region of Western Canada" (typescript of lecture), July 22, 1981, Agriculture Canada Research Station, Lethbridge; *Country Guide*, September 1955; *Annual Report, Department of Agriculture, Province of Saskatchewan, 1908*, 128–33.
23. Friesen, *The Canadian Prairies*, 33.
24. Symes, "The Evolution of Tillage Practices," 101.
25. John Bracken, *Dry Farming in Western Canada* (Winnipeg: Grain Growers' Grain Co., 1919), 111–12; [Evan Hardy] *Plows and Plowing* (Saskatoon: University of Saskatchewan, Agricultural Extension Bulletin No. 32, 1927), 2.
26. *The Nor'-West Farmer*, February 20, 1915.
27. Ibid., January 8, 1919.
28. Bracken, *Dry Farming in Western Canada,* 118.
29. Seager Wheeler, *Seager Wheeler's Book on Profitable Grain Growing* (Winnipeg: The Grain Growers' Grain Co., 1919), 86.
30. Ibid.; *The Western Producer*, March 7, 1929. Special equipment, including a twitch grass drag (twitch grass, couch grass, and quack grass were all synonymous), was sometimes used to loosen the roots which were then collected with a long-tooth drag harrow or a chain harrow (*The Nor'-West Farmer*, January 5, 1924).
31. Bracken, *Dry Farming in Western Canada,* 114.
32. Wheeler, *Profitable Grain Growing*, 185–86.
33. *The Nor'-West Farmer*, May 5, 1920.
34. Bracken, *Dry Farming in Western Canada,* 114.
35. Wheeler, *Profitable Grain Growing*, 188.

36. Morton, *Manitoba: A History*, 395.
37. Morton, "Agriculture in the Red River Colony," 314.
38. *The Grain Growers' Guide*, March 20, 1919.
39. Rogers and Anderson, *The History, Development and Use of Tillage and Cultivation Implements*, 188–90.
40. *The Nor'-West Farmer*, May 5, February 5, and December 21, 1914; *Canadian Farm Implements*, September 1909 and November 1908; *The Grain Growers' Guide*, August 21, 1909.
41. *Canadian Farm Implements*, August 1912; Saskatchewan Corporate Registry, 16, File 000978, SABR.
42. *The Nor'-West Farmer*, April 5, 1917, and March 20, 1918.
43. Ibid., May 15, 1914, April 20, 1915, and August 20, 1917.
44. Ibid., February 6, 1922, September 6, 1920.
45. *PFRA Annual Report, 1937*, "Agricultural Improvement Association" (typescript), A.E. Palmer Papers, M6880, file 28, GAI.
46. *The Nor'-West Farmer*, March 1938; *Annual Report, Dominion Experimental Station, Swift Current, 1937*, 188–97. One-ways were also modified to achieve the same purpose.
47. *Guide to Farm Practice in Saskatchewan 1933*, 50.
48. Symes, "Evolution of Tillage Practices," 101; A.E. Palmer, "Dry Land Tillage and Equipment: Recent Developments," XIII A24, ca. 1940, LRS. There was yet no common name for the machine; it was often simply known by the manufacturer's brand name.
49. Symes, "Evolution of Tillage Practices," 36.
50. *Annual Report, Dominion Experimental Station, Swift Current, 1931*, 85.
51. "The Pattern of Mechanization and Wartime Changes on Farms in the Elrose-Rosetown-Conquest Area"; *Country Guide*, September 1942; Hardy to Cory Bros. Ltd., January 29, 1951, Hardy Papers, MG575, Box 3, University of Saskatchewan Archives.
52. Symes, "Evolution of Tillage Practices," 103; *Maintenance and Operation of Tillage Machinery* (Saskatoon: University of Saskatchewan, College of Agriculture, Bulletin 32, 1953), 16; James Gray, *Men Against the Desert* (Saskatoon: Western Producer Prairie Books, 1967), 232.
53. H.A. Lewis, "The Development of the Discer in Saskatchewan" (typescript), H.A. Lewis, Gray, Saskatchewan, 4.
54. Ibid., 2.
55. Ibid., 5–6; Interview with Al McKinnon, 1990; *Annual Report, AMA, 1961–62*, 2.
56. *Canadian Farm Equipment Dealer*, May 1969.
57. Corbet, *The Manufacture of Agricultural Implements in Alberta*, 86–90.
58. *Country Guide*, September 1942; *Farm and Ranch Review*, November 1952; Various correspondence, C.S. Noble, 1937–38, M7198, files 19 and 21, GAI.
59. Corbet, *The Manufacture of Agricultural Implements in Alberta*, 87–90; Johnston, "History of Agriculture in the Prairie Region."
60. "The G.L. Dodds Implements" (catalogue), 1914 (National Museum of Science and Technology, Ottawa, AGR D6398 3002 c 1914).
61. *The Nor'-West Farmer*, June 20, 1919.
62. H. Chester, "The Development of Soil Drifting Control in Southern Alberta," Seminar Presentation, 1953, XX D18, LRS; Gray, *Men Against the Desert*, 229.
63. "Blade Cultivators For Ploughless Tillage," April 22, 1940 (typescript), XIII B1, LRS.
64. Corbet, *The Manufacture of Agricultural Implements in Alberta*, 129–30; Rogers and Anderson, *The History, Development and Use of Tillage and Cultivation Implements*, 229–32.
65. Rogers and Anderson, *The History, Development and Use of Tillage and Cultivation Implements*, 232–36; Corbet, *The Manufacture of Agricultural Implements in Alberta*, 68, 129–30.

66. Noble to Margetts, November 26, 1938, M7198, file 43, GAI. On his finances, see Noble to Mayhood, May 2, 1938, Ibid., file 53.
67. *Western Business and Industry*, October 1946.
68. *Sunny South News*, March 31, 1976.
69. Interview with Victor Erdman, June 9, 1990.
70. Margetts to Noble, December 5, 1938, M7198, File 38, GAI.
71. Rogers and Anderson, *The History, Development and Use of Tillage and Cultivation Implements*, 229. On protecting patents, see correspondence in Noble Papers, M7687, files 2–8, GAI. On material supply problems, see various correspondence 1937, M7198, file 7, GAI.
72. *Maintenance and Operation of Tillage Machines* (Bulletin 32), 32.
73. *Country Guide and Nor'West Farmer*, April 1938.
74. Arthur D. Little Inc., "A Survey of Farm Equipment Manufacturing Opportunities in Manitoba. Report to the Department of Industry and Commerce, Province of Manitoba, March 1966" (Manitoba Legislature Library), 36.
75. *The Western Producer*, March 5, 1953.
76. *Canadian Farm Implements*, August 1951 and June 1963.
77. *Annual Report, AMA, 1961–62*, 22.
78. *The Western Producer*, December 20, 1962
79. *Free Press Weekly*, August 7, 1963.
80. "Machinery for Tillage and Planting" (Edmonton: Alberta Agriculture, 1984). A lighter version of this implement was the field cultivator. It had higher, more closely spaced shanks. It was used for pre-seeding tillage but was not heavy enough for deep banding of fertilizer.
81. *Regina Leader-Post*, September 3, 1974, and June 28, 1976.
82. *Regina Leader-Post*, June 28, 1976.
83. *Regina Leader-Post*, September 3, 1974.
84. "The Morris Story," 5.
85. Interview with Terry Summach, February 27, 1991; Interview with Terry Friggstad, February 27, 1991.
86. W.H. Fairfield, "The Rod Cultivator" (Ottawa: Canada Experimental Farms, Pamphlet No. 28, 1919).
87. Rogers and Anderson, *The History, Development and Use of Tillage and Cultivation Implements*, 237, 245.
88. "Jackson Rotary Rod Weeder" (poster), 1928 (GAI Library); "Samson Power Take Off Rod Weeder" (pamphlet), ca. 1961, WDMS II–1; Corbet, *The Manufacture of Agricultural Implements in Alberta*, 117–23.
89. Interview with Adeleine Morris, May 30, 1990; Interview with Dick Deryk, June 1, 1990.
90. "The Morris Story," 8.
91. *Canadian Farm Implements*, August 1951.
92. Ibid., July 1939.
93. Ibid., November 1942 and March 1960.
94. *Canadian Farm Equipment Dealer*, July 1968.
95. Rogers and Anderson, *The History, Development and Use of Tillage and Cultivation Implements*, 239–40; *Canadian Farm Implements*, July 1961.
96. *Annual Report, AMA, 1963–64,* 330. There had been other, apparently unsuccessful, attempts to use a power-take-off drive rod *(Canadian Farm Implements,* August 1951).
97. *Canadian Farm Implements*, September 1940.
98. Rogers and Anderson, *The History, Development and Use of Tillage and Cultivation Implements*, 240.
99. *Regina Leader-Post*, March 8, 1961.

100. "Victory Self-Rotating Rod Weeder" (pamphlet) ca. 1965, II–4, WDMS.
101. *Canadian Farm Equipment Dealer*, September 1965; Edwards Rod Weeder Ltd, November 23, 1972, RG20, V2040 file V.8001–3710/M1, NAC; "Edwards Heavy Duty Cultivators" (pamphlet) ca. 1970, II–1, WDMS.
102. *Guide to Farm Practice in Saskatchewan 1933*, 48–50.
103. Irving to Noble, December 29, 1937, M7198, file 49, GAI; Rogers and Anderson, *The History, Development and Use of Tillage and Cultivation Implements*, 181.
104. Interview with Terry Summach, February 27, 1991.
105. *Country Guide*, March 1957.
106. *Canadian Farm Implements*, January 1955.
107. Ibid.
108. *Country Guide*, March 1957.
109. Interview with Terry Summach, February 27, 1991; *Saskatoon Star Phoenix*, September 8, 1979.
110. *Guide to Farm Practice in Saskatchewan 1933*, 49–50; *Canadian Farm Implements*, September 1936.
111. *Guide to Farm Practice in Saskatchewan 1948*, 26.
112. *The Story of CCIL* (pamphlet) (Winnipeg: CCIL, 1948) (GAI Library), 4–5; Interview with H.A. Lewis, May 30, 1990.
113. *The Story of CCIL*, 4–5; *Canadian Farm Implements*, February 1946; Report to CCIL Directors, 1961, P846, PAM.
114. "Crown 7–2–1 Harrow Weeder" (catalogue) ca. 1960.
115. *Winnipeg Tribune*, May 16, 1979.
116. *Plows and Plowing* (University of Saskatchewan, Bulletin No. 32, 1927) 2; See also *The Nor'-West Farmer*, September 21, 1931.
117. *The Nor'-West Farmer*, July 5 and 20, 1932.

Seeding, Fertilizing, and Applying Chemicals *(pages 139–57)*

1. Russell Anderson, "Grain Drills Through Thirty-nine Centuries," *Agricultural History* 10 (1936): 157–205; Lillian Church, "History of Grain Drills," U.S. Department of Agriculture, Bureau of Agricultural Engineering, Information Series No. 70 (October 1935); *Canadian Farm Equipment Dealer*, March 1970.
2. *The Nor'-West Farmer*, May 1891; Morton, *Manitoba: A History*, 255; E.A. Docking, *Seeding Machinery* (Saskatoon: Extension Division, University of Saskatchewan, Publication No. 213, 1970), 7–8; Bracken, *Dry Farming in Western Canada*, 120.
3. *Saskatchewan Guide to Farm Practices, 1933*, 51.
4. "Noble Drill" (pamphlet), n.d. ("Noble Cultivators" file, GAI library).
5. D.T. Anderson, *A History of Seeding Practices and Machines in Alberta Agriculture, 1870–1955* (Wetaskiwin: Reynolds Alberta Museum, Background Paper 29, 1985), 58.
6. Ibid., 58–59.
7. In 1937, C.S. Noble reported extensive use in southern Alberta of rod weeders attached to drills. Newly seeded fields were also commonly rod weeded (Noble to Featherstonaugh, November 10, 1937, M7198, file 24, GAI).
8. Docking, *Seeding Machinery*, 9; *Canadian Farm Implements*, August 1963.
9. *The Western Producer*, January 3, 1980.
10. Docking, *Seeding Machinery*, 7–8; *Guide to Farm Practice in Saskatchewan, 1969*, 38; *Prairie Farmers' Catalogue 1990*, 158; *Canadian Agricultural Machinery for World Markets* (Ottawa: Canada External Affairs, n.d.), 31.
11. Docking, *Seeding Machinery*, 5; Bracken, *Dry Farming in Western Canada*, 120.
12. For example, *Guide to Farm Practice, 1933*, 50–51; Duncan Marshall, *Field and Farm Yard* (Imperial Oil [Canada] Ltd., 1929), 211; *Canadian Farm Implements*, March 1950.
13. Wheeler, *Profitable Grain Growing*, 189–91.

14. *Guide to Farm Practice in Saskatchewan 1948*, 29; see also *The Western Producer*, March 7, 1929.
15. *The Nor'-West Farmer*, March 5, 1924.
16. *The Nor'-West Farmer*, March 5, 1929.
17. "Kirchner's 1930 Seeder Plows" (pamphlet), Ag 12 II, Group B, file 39, SABS.
18. *Annual Report, Dominion Experimental Station, Swift Current, 1931*, 84.
19. *Guide to Farm Practice in Saskatchewan, 1963*, 38.
20. Docking, *Seeding Machinery*, 3; *Canadian Farm Equipment Dealer*, March 1970; *Guide to Farm Practice in Saskatchewan, 1963*, 38.
21. Docking, *Seeding Machinery*, 5; *Discers* (Ottawa: Agriculture Canada Publication 1569; 1976), 25–26.
22. *The Nor'-West Farmer*, July 5 and 20, 1932.
23. *Prairie Farmers Catalogue, 1990*, 158.
24. Corbet, *The Manufacture of Agricultural Implements in Alberta*, 130; *Annual Report, Dominion Experimental Station, Swift Current, 1952*, 38–42; D.T. Anderson, *A History of Seeding Practices*, 56–57; Docking, *Seeding Machinery*, 10–11.
25. *Regina Leader-Post*, July 2, 1977.
26. *The Western Producer*, July 7, 1977.
27. Ibid., June 4, 1981.
28. Ibid.
29. Interview with Terry Summach, February 27, 1991; Interview with Terry Friggstad, February 27, 1991.
30. D.G. Wetherell and Associates Ltd., *Mixed Farmyards in Alberta, 1895–1955* (Wetaskiwin: Reynolds Alberta Museum, Background Paper 15, 1984), 103–7.
31. "Cleaning Grain" File, Provincial Museum of Alberta; J. MacGregor Smith, "Report on Grain Cleaning Attachments . . . ," 1930, MacGregor Smith Papers, 66.14/10, PAA.
32. Wheeler, *Profitable Grain Growing*, 248. See also *The Nor'-West Farmer*, April 6, 1905, and February 5, 1915; "Recommendations of the Farm Machinery Committee (Saskatchewan) and Dominion Agronomists."
33. Wetherell, *Farmyards*, 107–10; *The Nor'-West Farmer*, February 20, 1925.
34. *Canadian Farm Implements*, May 1919.
35. Wetherell, *Farmyards*, 111–12.
36. Ibid., 112–13.
37. On such equipment, see, for examples, *Farmers Advocate*, December 20, 1902, and July 13, 1904; *Canadian Farm Implements*, April 1908 and April 1909; *The Nor'-West Farmer*, February 20, 1914, January 5, 1916, May 21, 1917, and February 5, 1918; "Metal Specialty Co. Ltd.," Ag 12 II, Group B, file 53, SABS.
38. See, for example, *Canadian Farm Implements*, November 1912.
39. *Canadian Farm Implements*, January 1953.
40. *Seed Cleaning in Saskatchewan* (Regina: Saskatchewan Department of Agriculture, 1956), 44; *Canadian Farm Implements*, October 1937; "Forever Industries Ltd" (pamphlet) ca. 1937, X–1, WDMS; *Catalogue*, Provincial Exhibition of Manitoba, Brandon, 1945; *The Financial Post*, September 24, 1960; *Prairie Farmers' Catalogue, 1990*, 97.
41. Wayne D. Rasmussen, "The Impact of Technological Changes on American Agriculture, 1862–1962," *Journal of Economic History* 22 (1962):578.
42. D.G. Wetherell and Associates Ltd., *The Use of Fertilizers and Chemical Insecticides and Herbicides in Alberta Farming 1895–1960* (Wetaskiwin: Reynolds Alberta Museum, Background Paper 31, 1985), 12–17.
43. Ibid., 25–33.
44. *Annual Report, Dominion Experimental Station, Swift Current, 1937*, 165–69; *Grasshopper Bait Spreaders* (Regina: Saskatchewan Department of Agriculture, 1949). Chlordane, used for grasshopper control, came on the market in 1949. Baits made a partial

come-back in the 1970s and 1980s since they were more economical and were safer for bees than were sprays. They were applied either in granular form with granular applicators (also used for fertilizer) or by specially manufactured bait spreaders. By the 1980s, some of these had adapted pneumatic technology to spread the bait with a blower (*Canadian Farmers' Catalogue, 1990*, 46).

45. Wetherell, *The Use of Fertilizers and Chemical Insecticides and Herbicides*, 12–17.
46. *Canadian Farm Implements*, June 1946.
47. *Calgary Herald*, January 30, 1965.
48. *Canadian Farm Equipment Dealer*, January 1970.
49. For examples, see *Prairie Farmers' Catalogue 1990*, 39, 44, 45.
50. *Canadian Farm Implements*, March 1960 and May 1963. See also *Canadian Farm Implements*, March 1951; *Canadian Agricultural Machinery for World Markets*, 83.
51. *Canadian Farm Implements*, May 1962.
52. *Prairie Farmers' Catalogue 1990*, 39, 45, 44.
53. *Canadian Farm Implements*, May 1962.
54. Interview with Clem Roles, May 20, 1990.
55. Docking, *Seeding Machinery*, 5.
56. Wetherell, *The Use of Fertilizers and Chemical Insecticides and Herbicides*, 65–74; *The Country Guide*, March 1956.
57. Spierco Industries, November 23, 1972, RG20, V2040 V8001–3710/M1, NAC.

Harvesting Equipment *(pages 158–86)*

1. A cradle was a scythe equipped with prongs that held the grain as it was cut. It cut grain more quickly than the scythe or the sickle.
2. *Farm and Ranch Review*, April 1931; *Country Guide*, August 1959.
3. *The Nor'-West Farmer*, July 5 and 20, 1932.
4. *The Nor'-West Farmer*, July 5 and 20, 1932; Symes, "Agricultural Technology and Changing Life on the Prairies," 39.
5. Hamilton to Motherwell, February 22, 1923, M13, SABR.
6. *Canadian Farm Implements*, November 1908 and August 1909; Corbet, *The Manufacture of Agricultural Implements in Alberta*, 33–37.
7. *The Nor'-West Farmer*, September 5, 1918. On one case of farmers' losses from investing in stooking machine companies, see *The Grain Growers' Guide*, November 3 and December 1 and 15, 1915.
8. For example, *The Nor'-West Farmer*, September 5, 1918, and May 20, 1926. Binders too were difficult to use in lodged grain because the straw was often tangled and would not bind evenly.
9. *Canadian Farm Implements*, August 1912; *The Nor'-West Farmer*, May 20 and July 20, 1915, June 5, 1928; *The Grain Growers' Guide*, July 2, 1919; Stewart Sheaf Loader Co., Defunct Company Files, File 3304, SABR; "Announcing the Acme Jubilee Tractorized Loader" (poster), ca. 1928, VI–I, WDMS.
10. *Country Guide*, July 1947.
11. Ibid., August 1959.
12. O.S. Longman, "Pioneers and Pioneering in Western Canada" (unpublished mss.), 1963, pp. 52–54, O.S. Longman Papers, M4049, GAI; *The Nor'-West Farmer*, January 25, 1919.
13. Shepard, "The Mechanized Agricultural Frontier," 9–14.
14. For example, *Farmer's Advocate*, December 21, 1903, and December 14, 1904.
15. Don Gayton, "Threshing Days," *Saskatchewan Heritage Review*, 5 (1988):5.
16. Irwin, "The Sylvester Auto-Thresher," 143–50; Shepard, "Tractors and Combines in the Second Stage of Agricultural Mechanization," 262; L.H. Thomas, "Early Combines in Saskatchewan," *Saskatchewan History*, 8 (1955): 1–5; *The Nor'-West Farmer*, July 5 and 20, 1932.

17. *The Nor'-West Farmer*, March 20, 1916.
18. Evan Hardy, "The Combine Harvester in Western Canada," *Scientific Agriculture* 12 (1931):122.
19. "Field Crops Seminar, 1932," M2369, file 818A, GAI.
20. *The Nor'-West Farmer*, September 5, 1923.
21. Ibid., July 5 and 20, 1932.
22. Shepard, "Tractors and Combines in the Second Stage of Agricultural Mechanization," 263.
23. *Guide to Farm Practice in Saskatchewan, 1933*, 52; Hardy, "The Combine Harvester in Western Canada," 123.
24. Shepard, "Tractors and Combines in the Second Stage of Agricultural Mechanization," 265.
25. J.G. Taggart and J.K. Mackenzie, *Seven Years' Experience with the Combined Reaper Thresher 1922–1928* (Ottawa: Department of Agriculture, Bulletin 118, New Series, 1929), 3, 6.
26. "Field Crops Seminar, 1932"; Symes, "Agricultural Technology and Changing Life on the Prairies," 40.
27. Hanson, "History of Swathing and Swath Threshing," 1–5.
28. Sources for the Hanson swather and combine are: Hanson, "History of the Swather"; George Shepherd, "The Romance of the Grain Swather"; and "History of Swathing and Swath Threshing"; all in RG33/91, Vol. 21 file 11–2–1 Vol. 8, NAC.
29. Taggart and Mackenzie, *Seven Years' Experience*, 22.
30. *The Nor'-West Farmer*, August 20, 1929.
31. Thomas Isern, "The Header Stack-Barge: Folk Technology on the North American Plains," *The Social Science Journal* 24 (1987): 363; *The Header and Header Barge for Grain Harvesting* (Saskatoon: University of Saskatchewan, Agricultural Extension Bulletin No. 85, 1938), 11. Headers were often called "push-type binders."
32. Isern, "The Header Stack-Barge: Folk Technology," 363–66.
33. Dallas C. Wood, *Grain Cutting Implements in Alberta, 1870–1950* (Wetaskiwin: Reynolds Alberta Museum, Background Paper 7, 1983), 32.
34. Isern, "The Header Stack-Barge: Folk Technology," 367–70. See also *Annual Report, Dominion Experimental Station, Swift Current, 1933*, 44.
35. Isern, "The Header Stack-Barge: Folk Technology," 367, 370.
36. *The Nor'-West Farmer*, September 5, 1923.
37. "Gregg Harvester Stacker" (pamphlet), ca. 1931, VI–I, WDMS; *Canadian Farm Implements,* February 1928; Hardy, "The Combine Harvester in Western Canada," 127.
38. "Gilbert Harvester Stacker and Side Delivery Feeder" (pamphlet), 1937, and Gilbert to Manning, October 17, 1936, 73.307/378, PAA.
39. *The Western Producer,* May 14, 1987.
40. Symes, "Agricultural Technology and Changing Life on the Prairies," 40; *Canadian Farm Implements*, May 1951; *Regina Leader-Post*, March 14, 1957; *Self Propelled or Pull Type Combine?* (Edmonton: Alberta Agriculture Farm Business Management Board, 1980), 4.
41. "The Pattern of Mechanization and Wartime Changes on Farms in the Elrose-Rosetown-Conquest Area."
42. *Country Guide*, May 1955.
43. "Versatile Manufacturing Co. Western Canada's Largest Farm Implement Manufacturer, Our History" (Winnipeg: Versatile Ltd., pamphlet), ca. 1971; "Ford New Holland" (catalogue), 1987; *Winnipeg Tribune,* February 16, 1967; *Manitoba Business,* August 1984.
44. G. Quick and W. Buchele, *The Grain Harvesters* (St Joseph, Michigan: The American Society of Agricultural Engineers, 1978), 245–56; *Canadian Farm Equipment Dealer*, May 1969.

45. "The Cambrian Group Feasibility Study. Report to Saskatchewan Agricultural Implements Board, May 1972," R622, file 7, SABR; *Regina Leader-Post*, November 30, 1971.
46. Interview with W.B. Reed, May 17, 1990.
47. Ibid.; *Financial Post*, June 30, 1973.
48. H.L. Patterson, *The Farm Machinery Outlook in the Prairie Provinces* (Ottawa: Department of Agriculture Processed Publication, January 1945), 7; *Calgary Herald*, October 2, 1943.
49. *Country Guide*, July 1943; Clash to Longman, June 15, 1944, 73.307/373, PAA.
50. *Canadian Farm Implements*, February 1946.
51. *Winnipeg Tribune*, August 27 and 30, 1949; Hanson, "History of the Swather," n.p.
52. Memorandum, June 8, 1967, RG33/91, Vol. 4 1–7–2, NAC; "Killbery Industries," and Record of Interview, November 20, 1972, RG20, Vol. 2040 file V8001–3710/M1, NAC; *Canadian Farm Implements*, February 1953.
53. "Versatile Manufacturing, Our History," n.p.
54. Hanson, "History of the Swather"; *Winnipeg Tribune*, February 16, 1967; "Versatile Manufacturing, Our History."
55. *Canadian Agricultural Equipment and Machinery for World Markets*, 47.
56. *Canadian Farm Implements*, February 1953.
57. *Canadian Farm Implements*, July 1961.
58. *Canadian Farm Implements*, December 1946.
59. On corn equipment in Canada, see *Canadian Farm Implements*, April and March 1946.
60. Memorandum Re: Sugar Beet Machinery . . . and Potato Machinery," n.d., XIII A16, and file B17 "Basic Principles . . . Harvesting Sugar Beets," 1954, LRS.
61. *Lethbridge Herald*, March 20, 1982; Interview with Harold Kirchner, June 9, 1990.
62. Corbet, *The Manufacture of Agricultural Implements in Alberta*, 69–70.
63. In 1942–43, Robinson Machinery Co., Calgary, brought out a beet loader designed by Briosi. Although largely experimental and poorly tested, wartime labour shortages led beet growers in southern Alberta to welcome the machines. Robinson's production of this equipment had ceased by about 1944 (Test Report, ca. 1944, XIII B5, LRS; Andrew Briosi, *Yesterday's Dropout*, P19851143004, p. 25ff, Sir Alexander Galt Museum and Archives).
64. *Regina Leader-Post*, September 5, 1952.
65. For examples, see *The Nor'-West Farmer*, July 15 and 20, 1915; *The Grain Growers' Guide*, December 29, 1909.
66. *Canadian Farm Equipment Dealer*, November 1968.
67. For examples, see *The Nor'-West Farmer*, August 21, 1905, August 20, 1917, and June 30, 1923.
68. *The Nor'-West Farmer*, June 5, 1917.
69. *Canadian Farm Implements*, August 1951.
70. *The Nor'-West Farmer*, November 5, 1923.
71. Interview with H.A. Lewis, May 30, 1990.
72. Interview with Peter Pakosh, October 3, 1990.
73. Corbet, *The Manufacture of Agricultural Implements in Alberta*, 41; *Canadian Farm Implements*, August 1951 and September 1956.
74. Dealer Notice, Sakundiak Farm Equipment, May 12, 1966; Interview with Peter Sakundiak, May 25, 1990.
75. Interview with John Buhler, September 30, 1990; Interview with Peter Bergen, May 23, 1990.
76. Questionnaire, Brandt Industries, 1990; "Imperial Steel Products," 1972, RG20, V2040 f V8001.37101M1, NAC.
77. Taggart and Mackenzie, *Seven Years' Experience*, 14.

78. "Company Profile, Labtronics" (typescript), 1990.
79. *Guide to Farm Practice in Saskatchewan, 1948*, 33.
80. See, for example, "Grain Dryer," ca. 1941, 73.307/373, PAA; "Farm Grain Dryers" (University of Alberta, pamphlet, ca. 1952).
81. *Guide to Farm Practice in Saskatchewan, 1963*, 41.
82. *PrairiFarmersssssCataloguł989*, 170; *Canadian Farm Equipment Dealer*, November 1968.
83. *The Western Producer*, October 19, 1972; *Edmonton Journal*, November 27, 1991; Questionnaires, 1991.

Hay and Livestock Equipment *(pages 187–99)*

1. *Farming in Canada* (Ottawa: Department of Agriculture, 1950), 53; Patterson, *The Farm Machinery Outlook*, 26.
2. While there were a few horse ranches before 1925 and a few sheep ranches before World War II, neither were extensive.
3. Johnston, "History of Agriculture in the Prairie Region," 3.
4. "Saskatchewan" (typescript), ca. 1917, R266, file 51a, SABR.
5. Spector, *Agriculture on the Prairies*, 5.
6. Johnston, "History of Agriculture in the Prairie Region," 3.
7. *Farm and Ranch Review*, January 1957.
8. "The Wonder Mower" (pamphlet), 1922, VII–1, WDMS.
9. *The Nor'-West Farmer*, June 20, 1916. On the Prairies, the term windrow was normally reserved for hay, while the term swath referred to grain.
10. Ibid., July 5, 1919.
11. Although it led to high spoilage, in some areas hay was simply gathered with a sweep and left in the field in large piles. In other cases, it was stored in barn lofts, which by World War I had been mechanized with large forks or slings which elevated the hay and carried it into the loft along a track fixed to the ceiling of the loft.
12. *Annual Report, Dominion Experimental Station, Swift Current, 1945*, 230.
13. Ibid., 1944, 248.
14. F.F. Parkinson, *Stook Sweeps* (Edmonton: Alberta Department of Agriculture, Circular No. 64, 1943).
15. Patterson, *The Farm Machinery Outlook*, 23; *Annual Report, Dominion Experimental Station, Swift Current, 1949*, 58–64.
16. *Country Guide*, April 1949.
17. Interview with Harold Kirchner, June 9, 1990; Corbet, *The Manufacture of Agricultural Implements in Alberta*, 68–69; *Guide to Farm Practice in Saskatchewan, 1972*, 18.
18. Report of Meeting, ASAE, December 6–8, 1943, RG17, 3216/155–2, NAC; *Country Guide*, September 1942; *Annual Report, Dominion Experimental Station, Swift Current, 1949*, 58–64.
19. "Survey of Equipment Being Used to Save Straw After Combining, 1948," Papers of the Alberta Department of Agriculture, 90.926, GSE, PAA.
20. Report on the AMA, 1964, p. 5, R372, fII–6, SABR.
21. "Memorandum on Farm Machines. Their Manufacture, Distribution and Prices. Based on the Experience of the CCIL," 19–20.
22. "The Malco Pop-up Bale Loader" (pamphlet), n.d. (GAI library). Malco Manufacturing was a division of Malmgren Manufacturing which was affiliated with Allied Farm Equipment Manufacturing, Chicago.
23. *Red Deer Advocate*, June 6, 1973; "Williamson Bale Stooker" (catalogue), ca. 1970 (GAI library). Williamson designed a subsurface tillage blade in 1945 when living in southern Alberta.
24. *Canadian Farm Equipment Dealer*, May 1973.

25. Ibid., April 1968 and May 1973.
26. Heslop to Thue (attachments), January 9, 1991, PIMA files.
27. *Annual Report, Dominion Experimental Station, Swift Current, 1949*, 58–64.
28. *The Nor'-West Farmer*, April 20, 1915, July 5, 1916, July 5, 1922.
29. Heslop to Thue (attachments), January 9, 1991, PIMA files.
30. *The Nor'-West Farmer*, November 5, 1914.
31. Russel H. Anderson, "The Technical Ancestry of Grain-Milling Devices," *Agricultural History* 12 (1938): 256–58. The chopping action of ensilage and hay cutters is, of course, a different mechanical principle from grinding.
32. *Guide to Farm Practice in Saskatchewan, 1972*, 19.
33. *Canadian Farm Implements*, July 1929; *Western Business and Industries* 19 (Part II, 1945):57.
34. Interview with John Buhler, September 30, 1990.
35. *Small Business*, July/August 1987; Interview with John Buhler, September 30, 1990.
36. Ibid.
37. Questionnaire, Koenders, 1990.
38. *Agricultural and Industrial Progress in Canada*, November 1947; Frank Shefrin, "Canadian Farmers Mechanize," *Economic Annalist* 19(1949):101–8; H.R. Shaw and R.C. Gilstorf, "Mechanization in Canadian Agriculture," *Economic Annalist*, 24(1954): 5–9.
39. For examples, see *The Nor'-West Farmer*, October 5, 1923; Evan Hardy, "Progress in Agricultural Engineering in Western Canada," Conference Paper, 1947, Faculty Publications, University of Saskatchewan Archives. Among these short line firms were Louden Manufacturing, Guelph, Ontario, and Beatty Brothers Ltd.
40. *Canadian Farm Equipment Dealer*, April 1971.
41. *Guide to Farm Practice in Saskatchewan, 1972*, 21. Liquid manure was spread with hydraulic pumps from a tank on a truck or from trailer-type slurry and vacuum tanks.
42. *The Nor'-West Farmer*, March 6, 1922.
43. "Metal Specialty Company Ltd." (catalogue), I–1, 631.3 M5876 1918, WDMS.
44. *Canadian Farm Implements*, March 1951.
45. On the general impact of electricity on barns and livestock care, see *The Country Guide*, July 1947 and February 16, 1949; Subcommittee to Thomson, May 17, 1944, August Fisher Papers, M376, file 421, GAI; Report of Meeting, ASAE, December 6–8, 1943, RG17, 3216/155–2, NAC.

Conclusion *(pages 200-5)*

1. This brief overview of future trends draws upon implications that have emerged from our analysis of the prairie industry as well as views expressed by industry and farm participants and observers in the course of our research interviews and in reply to a questionnaire circulated to members of PIMA. A list of figures interviewed is given in the bibliography.
2. C. Wayne Lindwall, "Erosion Control in Tillage Systems," Paper, courtesy of author, n.d.
3. C. Wayne Lindwall, "Minimum Tillage in Continuous Cropping Rotations," Address, Expert Committee on Weeds, Edmonton, 1985.

Interviews

D. Tracy Anderson (retired), Lethbridge Research Station, Lethbridge
Peter Bergen, Bergen Industries Inc., Drake, Saskatchewan
John Buhler, John Buhler Inc., Winnipeg
Ray Bussiere, High-Line Manufacturing Inc./Laurier Manufacturing Ltd., Vonda, Saskatchewan
Jim Carnago, Schulte Industries Ltd., Englefeld, Saskatchewan
Eugene Demkiw, Ezee-On Manufacturing Ltd., Vegreville, Alberta
Dick Deryk, Yorkton
Francis, Lionel, and Donald Doepker, Doepker Industries Ltd., Annaheim, Saskatchewan
Walter Dressel, CCIL, Winnipeg
Victor Erdman, Victory Equipment Ltd., Lethbridge
Terry Friggstad, Flexi-Coil Ltd., Saskatoon
Ross Giles, Degelman Industries Ltd., Regina
Harold Kirchner, Kirchner Machine Ltd., Lethbridge
H.A. (Hal) Lewis (retired), University of Saskatchewan, Gray, Saskatchewan
A. McKinnon (retired), University of Saskatchewan, Regina
Ray Malinowski, Leon Ram Enterprises Inc., Yorkton
Adeleine Morris, Morris Industries Ltd., Yorkton
Peter Pakosh, Versatile Ltd., Winnipeg
Clem Roles, Smith-Roles Ltd,, Saskatoon
Peter Sakundiak, Sakundiak Equipment Ltd., Regina
Gavin Semple, Brandt Industries Ltd., Regina
Paul M. Soubry, VFEO–Ford New Holland Canada Ltd., Winnipeg
Terry Summach, Flexi-Coil Ltd., Saskatoon

Bibliography of Cited Secondary Sources

Ajao, A.O., and Barr, B.M. "Entrepreneurial Assessment of Government Programmes to Encourage Alberta's Foreign Trade." *Prairie Forum* 2 (1977).

Andal, M.E. "Use of Power Machinery in Bush Land Improvement in North Eastern Saskatchewan." *Economic Analyst* 17 (May 1947).

Anderson, D. Tracy. *A History of Seeding Practices and Machines in Alberta Agriculture, 1870–1955.* Wetaskiwin: Reynolds Alberta Museum, Background Paper 29, 1985.

Anderson, Russell. "Grain Drills Through Thirty-nine Centuries." *Agricultural History* 10 (1936).

Anderson, Russel H. "The Technical Ancestry of Grain-Milling Devices." *Agricultural History* 12 (1938).

Ankli, Robert E. "Farm Income on the Great Plains and Canadian Prairies 1920–1940." *Agriculture History* 51 (1977).

Ankli, Robert E.; Helsberg, H. Dan; and Thompson, John Herd. "The Adoption of the Gasoline Tractor in Western Canada." In *Canadian Papers in Rural History II*, edited by D. Akenson. Gananoque, Ontario: Langdale Press, 1980.

Archer, John. *Saskatchewan: A History.* Saskatoon: Western Producer Prairie Books, 1980.

Barr, Brenton M. "Reorganization of the Economy Since 1945." In *Studies in Canadian Geography*, edited by P.J. Smith. Toronto: University of Toronto Press, 1972.

Broehl, W.J. *John Deere's Company: A History of Deere and Company and Its Times.* Doubleday and Co. Inc., 1984.

Brook A.N., and Kampouris, P. *Statistics Relating to Farm Machinery in Canada 1950–1976.* Ottawa: Agriculture Canada, 1977.

Brown, Tom. "Canadian Agricultural Equipment Manufacturers Advertisement Index, 1847–1942." *Material History Bulletin* Spring 1987.

Carter, Sarah. *Lost Harvests: Prairie Indian Reserve Farmers and Government Policy.* Montreal: McGill–Queens University Press, 1990.

Church, Lillian. "History of Grain Drills." United States Department of Agriculture, Bureau of Agricultural Engineering, Information Series No. 70, October 1935.

Clark, W. Leland. "Harry Cater: The Personification of the Successful Municipal Politician?" *Historical and Scientific Society of Manitoba, Transactions* Series III (1977–79).

Co-operative Leadership: Harry Fowler's Memoirs. Introduction by T. Phalen. Saskatoon: Co-operative College of Canada, 1977.

Corbet, Elise A. *The Manufacture of Agricultural Implements in Alberta 1870 to 1955.* Wetaskiwin: Reynolds Alberta Museum, Background Paper 20, 1984.

"Costs of Clearing and Breaking Bush Lands in Alberta." *Economic Annalist* 14 (Feb 1944).

Dick, Lyle. *Farmers "Making Good" The Development of Abernathy District, Saskatchewan 1880–1920.* Ottawa: Parks Canada, 1989.

Edwards, Les. *An Idea and Fourteen Dollars.* Saskatoon: Smith-Roles Ltd, ca. 1980.

Fay, Terence J. *Farm Implement Firms as Symptomatic of the Rise of Regional Grain Cities: Winnipeg and Minneapolis, 1876–1926.* Winnipeg: Institute of Urban Studies, University of Winnipeg, 1980.

Fifty Years of Progress on Dominion Experimental Farms, 1886–1936. Ottawa: King's Printer, 1939.

Fowke, Vernon. "Economic Effects of the War on the Prairie Economy." *Canadian Journal of Economics and Political Science* XI (1945).

Friesen, Gerald. *The Canadian Prairies: A History.* Toronto: University of Toronto Press, 1984.

Gayton, Don. "Threshing Days." *Saskatchewan Heritage Review* 5 (1988).

Good, W.S., and Beckman, M.D. "A Prairie Distribution System in Transition: The Case of Farm Machinery Parts in Manitoba." *Prairie Forum* 2 (1977).

Gordon, Stanley. *Agricultural Tractors in Alberta Since 1925.* Wetaskiwin: Reynolds Alberta Museum, Background Paper 17, 1983.

Gray, James. *Men Against the Desert.* Saskatoon: Western Producer Prairie Books, 1967.

Grose, Rex E. "Manufacturing in Manitoba: Growth and Trends." *Canadian Geographical Journal,* March 1953.

Ham, Penny. *Manitoba Agricultural Museum.* Austin, Manitoba: Manitoba Agricultural Museum, 1979.

Hanson, Helmer H. "History of Swathing and Swath Threshing." RG33/91, Papers of the Royal Commission on Farm Machinery [Barber Commission], Vol. 21 file 11–2–1 Vol. 8, NAC.

"History of the Swather." in RG33/91, Papers of the Royal Commission on Farm Machinery [Barber Commission], Vol. 21 file 11–2–1 Vol. 8, NAC.

Irwin, Ross. "The Sylvester Auto-Thresher: A Canadian First in Grain Harvesting." *Scientia Canadensis* 8 (1984).

Isern, Thomas. "The Header Stack-Barge: Folk Technology on the North American Plains." *The Social Science Journal* 24 (1987).

Johnston, Alex. "History of Agriculture in the Prairie Region of Western Canada." Typescript of lecture, July 22, 1981, Agriculture Canada Research Station, Lethbridge.

Jones, Fred. *Farm Gas Engines and Tractors.* New York and London: McGraw-Hill Book Co., 1938.

Kristjanson, Baldur H. "The Capital Structure of Canadian Agriculture." *Economic Analyst* 27 (1957): 57–61 (Part I) and 101–5 (Part II).

Kulshrestha, S.N., and Devine, D. "The Crowsnest Rate and Prairie Development. 1. A Historical Review." Saskatoon: University of Saskatchewan, Agricultural Scene, Publication No. 298A, 1978.

Lewis, H.A. "The Development of the Discer in Saskatchewan." Typescript, H.A. Lewis, Gray, Saskatchewan, n.d.

Lindwall, C. Wayne. "Minimum Tillage in Continuous Cropping Rotations." Address, Expert Committee on Weeds, Edmonton, 1985.

——. "Erosion Control in Tillage Systems." Paper, courtesy of author, n.d.

Longman, O.S. "Pioneers and Pioneering in Western Canada." Unpublished mss., 1963, Longman Papers, M4049, GAI.

MacPherson, Ian. "'Better Tractors For Less Money': The Establishment of Canadian Co-operative Implements Ltd." *Manitoba History* 13 (1987).

Made in Saskatchewan: A Story of Invention. Index 1905–1980. Saskatoon: Western Development Museum, 1980.

Mahaffy, A.W. *The Machine Process in Agriculture With Special Reference to Western Canada.* Saskatoon: University of Saskatchewan, Department of Economics, 1923.

Martinusen, D., and Barry, B. *Revenues, Costs and Profits in the Farm Machinery Industry.* Ottawa: Royal Commission on Farm Machinery, 1970.

Mechanization and Farm Costs. Royal Commission on Agriculture and Rural Life. Regina: Queen's Printer, 1955.

"The Morris Story." Prepared by Western Construction and Industry Magazine in Conjunction with the Morris Rod Weeder Company 50th Anniversary, March 15, 1979.

Morrison, J.W. "Agricultural Achievements of the Prairie Experimental Farms 1886–1986." *Prairie Forum* 11 (1986).

Morton, W.L. "Agriculture in the Red River Colony." *Canadian Historical Review* 30 (1949).

——. *Manitoba: A History*. Toronto: University of Toronto Press, 1957.

Norrie, Kenneth H. "Agricultural Implement Tariffs, the National Policy and Income Redistribution in the Wheat Economy." *Canadian Journal of Economics* 7 (1974).

——. "Some Comments on Prairie Economic Alienation." *Canadian Public Policy* 2 (1976).

Patterson, H.L. *The Farm Machinery Outlook in the Prairie Provinces*. Ottawa: Department of Agriculture Processed Publication, January 1945.

Phillips, W.J. *The Agricultural Implement Industry. A Study in Competition*. Toronto: University of Toronto Press, 1956.

Proudfoot, Bruce. "Agriculture." In *Studies in Canadian Geography*, edited by P.J. Smith. Toronto: University of Toronto Press, 1972.

Quick, G., and Buchele, W. *The Grain Harvesters*. St Joseph, Michigan: The American Society of Agricultural Engineers, 1978.

Rasmussen, Wayne D. "The Impact of Technological Changes on American Agriculture, 1862–1962." *Journal of Economic History* 22 (1962).

Richards, J.H. "Retrospect and Prospect." In *Studies in Canadian Geography,* edited by P.J. Smith. Toronto: University of Toronto Press, 1972.

Richeson, Dave. "Farming Moves into High Gear." *Horizon Canada,* 1987.

Rogers, Barry, and Anderson, D. Tracy. *The History, Development and Use of Tillage and Cultivation Implements in Alberta Agriculture, 1870–1955*. Wetaskiwin: Reynolds Alberta Museum, Background Paper 26, 1985.

Schweitzer, Thomas T. *Old Myths and New Choices: Railway Freight Rates and Western Economic Development*. Ottawa: Economic Council of Canada, Discussion Paper No. 269, 1984.

Seifried, A.N. "The Westward Shift of Manufacturing on the Prairies." *Prairie Forum* 1 (1986).

Shaw, H.R., and Gilstorf, R.C. "Mechanization in Canadian Agriculture." *Economic Annalist* 24 (1954).

Shefrin, Frank. "Canadian Farmers Mechanize." *The Economic Annalist* 19 (1949).

Shepard, R. Bruce. "The Mechanized Agricultural Frontier of the Canadian Plains." *Material History Bulletin*, Spring 1979.

——. "Tractors and Combines in the Second Stage of Agricultural Mechanization on the Canadian Plains." *Prairie Forum* 11 (1986).

Shepherd, George, "The Romance of the Grain Swather." Typescript in RG33/91, Vol. 21 file 11–2–1 Vol. 8, NAC.

Simkin, Rubin. *The Prairie Farm Machinery Co-operative: The "CCIL."* Ottawa: Queen's Printer, 1970.

Sonntag, B.H. "Farm Machinery Syndicates in the Prairie Provinces." *Canadian Farm Economics*, August 1966.

Spector, David. *Agriculture on the Prairies*. Ottawa: Parks Canada, 1983.

Stewart, A. "The Report of the Special Committee on Farm Implement Prices 1937." *Canadian Journal of Economics and Political Science* 4 (1937).

The Story of CCIL (pamphlet), Winnipeg: CCIL, 1948.

Symes, O.L. "Agricultural Technology and Changing Life on the Prairies." *Seminar Pro-*

ceedings, Development of Agriculture on the Prairies. Regina: University of Regina, Canadian Plains Area Studies, 1975.

——. "Evolution of Tillage Practices for Sustained Crop Production in the Palliser Triangle." *Transactions of the Royal Society of Canada* Series V/Vol. 1 (1986).

——. "Chronology of Agricultural Machinery Development and Related Information" in *Seminar Proceedings, Development of Agriculture on the Prairies.* Regina: University of Regina, Canadian Plains Area Studies, 1975.

Thomas, L.H. "Early Combines in Saskatchewan." *Saskatchewan History* 8 (1955).

Thompson, John Herd. *The Harvests of War: The Prairie West 1914–1918.* Toronto: McClelland and Stewart, 1978.

Thue, Ivan. *The PIMA Story.* Regina: Prairie Implement Manufacturers Association, 1990.

Tingley, K. *The Gasoline Tractor in Western Canada: An Introduction to its History in Alberta Before 1925.* Wetaskiwin: Reynolds Alberta Museum, Background Paper 18, 1984.

Tyrchniewicz, E.N. *An Exploratory Analysis of Transport Problems Relating to the Supply of Farm Machinery and Repair Parts to Prairie Farmers.* Ottawa: Canadian Transport Commission, 1973.

"Versatile History." Typescript ca. 1981, courtesy Peter Pakosh.

Versatile Manufacturing Co. Western Canada's Largest Farm Implement Manufacturer, Our History (pamphlet). Winnipeg: Versatile Ltd., ca. 1971.

D.G. Wetherell and Associates Ltd. *Mixed Farmyards in Alberta, 1895–1955.* Wetaskiwin: Reynolds Alberta Museum, Background Paper 15, 1984.

——. *The Use of Fertilizers and Chemical Insecticides and Herbicides in Alberta Farming 1895–1960.* Wetaskiwin: Reynolds Alberta Museum, Background Paper 31, 1985.

Wood, Dallas C. *Grain Cutting Implements in Alberta, 1870–1950.* Wetaskiwin: Reynolds Alberta Museum, Background Paper 7, 1983.

J.D. Woods and Gordon Ltd. *The Canadian Agricultural Machinery Industry.* Ottawa: Royal Commission on Canada's Economic Prospects, 1956.

Regular Members of the Prairie Implement Manufacturers Association 1989–1992

and those companies that completed a questionnaire on the history of their businesses

409904 Alberta Incorporated, Torrington, Alberta (Dennis G. Oster; Kathryn L. Oster). Manufactured "Liner Tamers" used to remove old inflations from milking units and to replace new inflation into the steel shell.

A & R Industries Ltd., Vonda, Saskatchewan (Philippe F. Denis, Pres.; Bernard J. Denis, Sec.). A farm steel fabricator.

Acrylon Plastics MB (1983) Inc., Winkler, Manitoba (Ted Northam, Vice-pres.; Marvyn Friesen, Gen. Mgr.) Manufactured rotational molded and vacuum-formed plastic products.

Adams Grain Dryer Co. Ltd., Saskatoon, Saskatchewan (Vic Adams). Manufactured grain dryers and heated aeration systems.

ADJA Industries Ltd., Calgary, Alberta (Dennis Poffenroth, Pres.). Established in 1977 to produce field and turf marker systems patented by the company. Later, sprayers and liquid applicators were added to the product line and an assembly branch was opened in Salt Lake City. Export markets were Australia, Europe, South Africa, and the USA.

AG Belting Inc., Regina, Saskatchewan (John M. Verbeek, Pres. & Gen. Mgr.; Keith Belanger, Cust. Serv.; Jack Powless, Plant Mgr.). In 1977 John Verbeek developed a method of welding cleats to belting and went into production, initially in Ontario. As most sales were in western Canada, the company moved to Regina in 1986. The company manufactured a complete line of agricultural conveyor belts as well as several other products and its export markets were in Australia, Uruguay, and the USA.

Ag Shield Manufacturing, Benito, Manitoba (Gary McCrea; Tom McCrea). Incorporated as McCrea Farms Ltd. in 1975, the company manufactured shielded sprayer booms, fertilizer carts, and autowater systems for horses, all of which were invented by the owners. Export markets were in Germany and the USA.

A.G.M. Steel Industries Ltd., Winfield, B.C. (Dwayne L. Armeneau, Pres.; Lee V. Wright, Plant Mgr.). Manufactured truck and bus parts, tanks, and fork lift attachments.

Airmaster Sales Ltd. (Emerson South), Winnipeg, Manitoba (David B. McGibbon, Agric. Mgr.; Ken MacCharles, Pres.). The company was in the industrial air movement business and established the agricultural division in 1979. It adapted existing technology to manufacture aeration fans, and later purchased another company and added sprayers and chaff spreaders to its product line. It exported to the USA.

Alberta Ag-Industries Ltd., Westlock, Alberta (Dwight Smith-Gander; Victor Julyan). Produced silage and grain bagging machines and a portable warehousing system using plastic bags.

AMS Industries Ltd., Vancouver, B.C. (Donald A. English; George Fry). Manufactured grain bin parts, blower inlets and shrouds, and tank ends.

Apollo Machine & Products Ltd., Saskatoon, Saskatchewan (Roy Swanson, Pres.; Don Swanson, Asst. Mgr.) Manufactured swath attachments, grain roller mills, snow blowers, and gravel and truck boxes.

Arc Alloy Welding (1984), Colonsay, Saskatchewan (Bill Stirton; Peggy Stirton). Manufactured trailers, bale feeders and spears, and "Calf Roper" chutes, and did custom manufacturing.

Armco Holdings Co. Ltd., Sceptre, Saskatchewan (John Armstrong, Pres.; Resa Armstrong, Sec. Treas.). Established in 1977 to manufacture trailers, silage baggers, and silage trucks, all of which were designed by the company and which incorporated technical improvements over the years.

Artex Fabricators Ltd., Surrey, B.C. (Ralph Terpstra; Henry Terpstra). The company started in 1971 with repair work, and as the business grew, it began to manufacture stall dividers and self-locking headrails. Most of the products, including feed and forage wagons, truck boxes, and livestock handling equipment, were designed and developed to meet the needs of the local dairy farmers. Their market was concentrated in British Columbia and the state of Washington.

Austrak Machinery Corporation, Weyburn, Saskatchewan (John Sprovieri, Pres.; Reg Carr, Vice-pres.). Austrak International was established in Melbourne, Australia, in 1975 and the Canadian affiliate was established in 1989. Products included the Austrak tractor, rotary brush cutter, front-end loader, and ground ripper.

Australian Canadian Agricultural Machinery Corp., Coaldale, Alberta (Jason Abrey; D.F. Hughes). This firm was established in 1980 and is a subsidiary of the Australian Agricultural Machinery Group originally incorporated in 1953. The Coaldale company manufactured hi-tech computer sprayers and exported to the USA.

Autotran Manufacturing Ltd., Loreburn, Saskatchewan (Frank Dietz; Brian Dietz). Manufactured swathers and did custom manufacturing.

Baker Electronic Enterprises Inc., Edmonton, Alberta (Kevin Robertson, Ag. Div. Mgr.; R. Ian Playfair, Sales Mgr.; Gordon Baker, Plastics Mgr.). Manufactured electronic fencers and monitors and did plastic mold manufacturing.

Baumle Enterprises Ltd., Kelsey, Alberta (Marvin Baumle, Pres.). Established in 1975, the company manufactured banding kits and pulse crop lifters.

BCM Manufacturing Ltd., Calgary, Alberta (Pierre Moreau). The company made plastic flooring for the livestock industry.

Bearing Buddies Inc., Regina, Saskatchewan (William Yarnton; William Rouse; Bob Koroluk). Manufactured bearing covers, pillow blocks, flange designs, and other models.

Behlen Industries, Brandon, Manitoba (Gord Doerksen, Pres.; Larry Arnold, Mktg. Mgr.; Victor Holodryga, Prod. Mgr.). Produced pre-engineered steel buildings, grain bins, culverts, well cribbing, and plastic pipe.

Beline Manufacturing Co. Ltd., Kindersley, Saskatchewan (W. John Bourne, Pres.; Alan M. Bourne, Vice-pres.). Established in 1962, the company manufactured pesticide, fertilizer, and preservative applicators that were invented by W. John Bourne. Other products included computerized sprayer-nozzle checkers, and computer and electronic controls for applicators. Most of the products were exported to the USA.

Bergen Industries Inc., Drake, Saskatchewan (Ed Bergen). Originally established in 1955 by Peter and Anne Bergen, the company was later managed by their three sons. The first product was a grain auger drill transport designed by Peter Bergen, and later equipment included stock trailers, drill and swather transports, harrow drawbars, and rock diggers. Some products were exported to the USA.

Big Island Ranch Limited, Sherwood Park, Alberta (John W. Carson; Donna M. Carson). Manufactured the hydraulic "Elk Squeeze."

Big "O" Inc., Abbotsford, B.C. (Barry Nybo; Hank Grunsky). Manufactured drainage tile for agricultural use.

Bill Douglas Manufacturing Ltd., Saskatoon, Saskatchewan (Larry McMartin; Kjell Madsen). Established by Bill Hartsthorne-Douglas in 1952 and sold in 1977 to McMartin and Madsen. The company manufactured trailers, press brakes, and tank skids, and did custom fabrication.

Bonar Inc. Plastic Molding Division, Winnipeg, Manitoba (Ken McKenzie, Sales Mgr.; Wayne Anderson, Gen. Mgr.). Established in 1973 as Global Plastics (Manitoba) Ltd. by Ken McKenzie and sold to Bonar in 1982. The plant designed and built a large grain hopper, discharge spouts, and drill fill spouts. It patented a unique septic tank and also manufactured molded sprayers, nurse and storage tanks, pig balls, and calf hutches. Export markets were developed in Australia and the USA.

Brandt Industries Ltd., Regina, Saskatchewan (Gavin Semple, Pres. & Gen. Mgr.; Don Vass, Purch. Mgr.; Dave Fletcher, Sales Mgr.). Brandt Electric was originally established in 1932 by Abram Peter Brandt. The company was formed in 1938 and offered welding and machine shop services. The first agricultural product was a wooden grain conveyor, followed by an auger a few years later. The company stayed in the family until 1976 when Gavin Semple became the manager. New agricultural products and expanded facilities were added over the years and product lines included grain augers, field sprayers, and wind cones. The company exported to the USSR and the USA.

Brekmar Industries Ltd., Saskatoon, Saskatchewan (Don Markusa, Pres.; Loretta Markusa, Sec.). Manufactured tanks, well casings, and hatch coverings, and did custom fabrication.

Bridgeview Manufacturing, Gerald, Saskatchewan (Kevin Hruska; Janice Andrusiak). Established in 1984 to manufacture hopper cones, construction cranes, and skid foundations, all of which were designed by Kevin Hruska. He also designed a cattle squeeze and the company grew steadily. Export markets were developed in the USA.

Bundus Enterprizes Ltd., Craven, Saskatchewan (Harold Bundus; Faron Bundus). Established in 1979, the company produced tarps, trailers, bale carriers, and unrollers, all under the Bundus Enterprizes trademark.

Butte Manufacturing Ltd., Ogema, Saskatchewan (Raymond Johnson; Shirley Johnson). Established in 1985, the company manufactured hopper bin unloading augers and transfer utility augers that were invented and patented by Raymond Johnson. Export markets were developed in the USA.

Cairns Iron Works, Coronach, Saskatchewan (Tom Cairns; Owen Cairns). Manufactured cattle lick tanks, hopper cones, zero till drill attachments, and swather cutting bar attachments.

Canadian Agtech. Partners, Olds, Alberta (Eric Jensen, Pres.; Sandy Harvey, Mktg. Mgr.). Manufactured solar water pumps.

Cancade Company Limited, Brandon, Manitoba (Paul Cancade; Ralph Redfern, Gen. Mgr.). Established in 1946 by the present owner's father. Product lines were developed in-house to reflect customer demand and included farm truck grain boxes, hydraulic hoists and tarps, and tractor-mounted front-end loaders and dozer blades. The company gradually expanded its service from a purely local market to the other Prairie provinces and the USA.

Cantrel Ltd., Calgary, Alberta (Larry Meriam, Pres.; Bill Wallace, Sales Mgr.). Manufactured dump bodies and hoists, crew cabs, cutaway vans, service and standard toppers, trailers, chippers, and stump cutters.

Carman Welding & Repair, Roland, Manitoba (Tony Suderman; Margaret Suderman). Manufactured wagons, cattle trailers, grain bins, and feeders.

Carter Day Industries (Canada) Ltd., Winnipeg, Manitoba (Mel Prokopowich, Plant Supt.;

Ron Mehmel, Sales). Manufactured grain cleaning, sizing, and separating equipment, and milling and air pollution control equipment.

Cattlelac Cattle Handling System Ltd., Red Deer, Alberta (Reg Cox, Pres.). Manufactured cattle-handling equipment.

Chigwell Industries Ltd., Lacombe, Alberta (Ronald Hilsabeck; Doug Hilsabeck). The company started in 1970 and manufactured wood and hopper fertilizer bins. The company grew steadily and by 1981 bins were being made of steel, and steel hopper cones were also manufactured.

Clarke Manufacturing Ltd., Rosetown, Saskatchewan (James A. Clarke; Murray A. Clarke). The company was formed in 1982 to manufacture "Hurricane" chaff spreaders, invented and patented by James A. Clarke. A new product line was bin ladders for steel grain bins.

CM Manufacturing Ltd., Armstrong, B.C. (Ken Topham; Jan Topham). Incorporated in Alberta in 1972 and in B.C. in 1987 to manufacture custom-built horse, stock, and utility trailers. The company marketed mainly in B.C., but also sold in the other Prairie provinces and exported to the USA.

Coaldale Iron Products Ltd., Coaldale, Alberta (Walter Isaak, Pres.; Jim Isaak, Gen. Mgr.). Established in 1975 and manufactured a rod weeder attachment, hydraulic hoses, and ornamental railings and gates, and did custom manufacturing and repairs. An export market was developed in the USA.

Commercial Industrial Mfg. Ltd., Humboldt, Saskatchewan (Henry Hauber; Mark Seidel). Incorporated in 1985 to custom manufacture steel products. Product lines included steel grain boxes, tarps, hopper bottoms, gravel boxes, and steel specialty items. Some products were exported to the USA.

Conservation Energy Systems, Saskatoon, Saskatchewan (Richard Olmstead, Pres.; Doug Wicks, Op. Mgr.). Manufactured heat recovery ventilation and modular integrated HVAC systems.

Cool Grain Inc., Calgary, Alberta (David Crompton, Pres.; Ron Larson, Prod. Mgr.). Manufactured grain temperature monitoring systems.

Courtney Berg, Linden, Alberta (Jim Courtney; Murray Regehr). Manufactured steel grain boxes, truck hoists, bale handlers, and hydraulic pumps, valves, and cylinders.

Craig's Mfg. (1989) Ltd., Morinville, Alberta (Phil Van Brabant; Cory Van Brabant, Sec. Treas.). Manufactured root windrowers, breaking and reforestation plows, bale wagons, and lube vans.

Custom Built Ag Industries Ltd., Gravelbourg, Saskatchewan (Brad Harvey, Gen Mgr.). Established in 1985 and manufactured trailers, header, swather, and round bale transports, and land levellers, all designed in-house. The company maintained steady growth and developed an export market in the USA.

Cyll Enterprises Ltd., Domain, Manitoba (Rick Manness; Norm Manness; Cecil Manness). Cecil Manness invented and patented the SAF-T-FIL bin level indicators and monitoring systems in 1983 and established the company in 1984.

D & D Machine Works Ltd., Lethbridge, Alberta (Dennis Barbour; Gunther Schulz, Acct.). Manufactured hydraulic cylinders, irrigation accessories, pulleys, and bearing housings.

Daleco-Built Products, St. Brieux, Saskatchewan (Dale Laczko; Keith Roenspies). Manufactured wireless-flow control for grain trucks.

Dalziel Enterprises Ltd., White Fox, Saskatchewan (Gilbert Dalziel; Barbara Dalziel). Manufactured leaf cutter bee nest blocks and cell removers, round bale unrollers, and dehydration and compaction equipment.

Dandi Shelters Western Inc., Lethbridge, Alberta (Jim Creason). Manufactured portable structures.

Dean Smith Acres Ltd., Swift Current, Saskatchewan (Dean Smith; Rheanne Smith). Established in 1979 and manufactured fertilizer tanks, PTO kits, cultivators, farm shop equipment, and bin sweeps. All products were designed and manufactured by Dean Smith as need and demand arose.

Degelman Industries Ltd., Regina, Saskatchewan (W.J. Degelman, Pres.; Ross Giles, Gen. Mgr.). In 1963, W.J. Degelman invented a rock picker on the family farm and marketed it from there. Two years later he incorporated the company and moved to Regina. The company sustained continuous growth and product lines expanded to include dozer blades, harrows, deep tillage cultivators, and rock rakes, all invented and patented by W.J. Degelman. Export markets were developed in Australia, Central America, Israel, Mexico, and the USA.

Del-Air Systems Ltd., Humboldt, Saskatchewan (Robert Hawkins, Pres.; Lloyd Crawford, Vice-pres., Mktg.; Miles Jorgenson, Vice-pres., R & D; Gene Gerspacher, Vice-pres., Fin.). Dick van Ee and Larry Benning invented and patented an agricultural heat exchanger in 1980 and established the firm in 1981 to manufacture the product. Later, ventilation fans, air inlets, and molded plastic products, all invented in-house, were added to the product line. The company has maintained continuous growth and developed export markets in Japan, Korea, Scandinavia, and the USA.

Dika Industries Ltd., Rycroft, Alberta (Mike Dika, Pres.; Greg Dika, Sales Mgr.). Brothers Mike and Jerry Dika established Dika Construction Ltd. in 1955, a logging and road construction business. Over the years it developed a product line and manufactured equipment invented by Mike Dika, for both the agricultural and forestry industries. These included root pickers, rakes, and harrows, rock windrowers, and land rollers. Much of this was built on a custom basis to serve the needs of northern Alberta, but the company also exported to the USA.

Doepker Industries Ltd., Annaheim, Saskatchewan (Donald Doepker, Pres.; Lionel Doepker, Sec. Treas.). Doepker Brothers started as a garage and repair shop in 1950 and went into manufacturing when Francis Doepker invented a harrow drawbar. Other inventions included a rod weeder, a harrow packer, drill and swather carriers, farrowing pens, self-feeders for the hog industry, and various types of trailers. Later, the company concentrated on the manufacture of trailers, which it marketed across Canada.

Doty Enterprises Inc., Carlyle, Saskatchewan (John G. Doty, Pres.; Irene R. Doty, Sales Mgr.). Established in 1980, the company manufactured hydraulic soil samplers, engine hoists, hydraulic presses, and trailers, all of which were developed and designed by John Doty. The company established an export market in the USA.

Double T Equipment Manufacturing Ltd., Airdrie, Alberta (Thomas Thomas, Pres.; Rey Rawlins, Sales Rep.). The company manufactured material-handling systems, including mushroom growing equipment, and fertilizer storage and blending systems.

Douglas Manufacturing Inc., Winkler, Manitoba (Bill Paulsen, Gen. Mgr.; John Heide, Sales Co-or.). Manufactured rototillers, mowers and cutters, scrapers and blades, pump carts, and a sprayer.

Dutch Industries Ltd., Regina, Saskatchewan (Izaak Cruson, Pres.; Lionel Kambeitz, CEO; Norman West, Prod. Mgr.). Established by P.J. Cruson in 1953 as a repair shop, the company later went into manufacturing agricultural equipment including chemical cab filters, chem-ease, crop lifters, banding knives and air seeders. The company maintained steady growth and developed export markets in the USA.

Dutertre Manufacturing Inc., Dinsmore, Saskatchewan (Don Dutertre). Manufactured chemical-handling equipment.

Dyna-Fab Industries Ltd., Watson, Saskatchewan (Martin Draude, Gen. Mgr.; Blaine Frederick, Prod. Mgr.). Manufactured drill and swather transports, V-ditchers, and hog barn equipment.

Ecklund Drivethru Gates, Broadview, Saskatchewan (Gary Ecklund; Dianne Ecklund). Established in 1986, this company manufactured drive-through gates that were invented and patented by Gary Ecklund. In the first two years, the equipment was made in the farm shop and marketed privately. The company then established itself in Broadview, added irrigation pivot tower gates to its product line, and developed an export market in the USA.

Edwards Grain Guard, Lethbridge, Alberta (Jim Edwards, Pres.; Milo Stronsmoe, Compt.). Edwards Rod Weeder Ltd. was incorporated in 1965 by Lawrence Edwards, whose sons purchased the company in 1984. They acquired Grain Guard Manufacturing and Sales Ltd. of Edmonton in 1990 and combined the operations of both companies. Their products included hoe drills, livestock waterers, and natural air grain-drying systems. Products were sold throughout North America.

E-Kay Enterprises, Sedgewick, Alberta (George Ekman; Joan Ekman). Established in 1986, the company manufactured bin sweeps and a variety of auger attachments, all invented by George Ekman. Products were marketed in western Canada and the USA.

Elias Scale Manufacturing, North Battleford, Saskatchewan (Gordon Elias; Laverne Elias). B.T. Johnson invented a portable hydraulic livestock scale and manufactured it in his garage for several years. When he sold the patent rights to Gordon and Laverne Elias in 1985, they established a company to produce the product in larger quantities. The company steadily increased its product lines to include hanging and truck scales and developed an export market in the USA and the United Kingdom.

Elmer's Welding and Manufacturing Ltd., Altona, Manitoba (Elmer Friesen; Elaine Friesen). Established in 1978, the company manufactured row crop cultivators and sprayers. It marketed in the Prairie provinces and exported to the USA.

Enviro Industrial Technology Corp., Winterburn, Alberta (Larry Cunningham). This company's major product was peat moss equipment.

Evertight Fence Anchor, Gleichen, Alberta (Ed Hoff; W.J. Dixon). Manufactured guy anchors that could be screwed into and out of the ground for agricultural and recreational uses.

Ezee-On Manufacturing Ltd., Vegreville, Alberta (Eugene Demkiw, Gen. Mgr.). Originally established in 1949, the company's first product was a fanning mill invented by John Phillips, a local farmer. In 1966, an "easy-on easy-off" front-end loader was patented and the firm's name was changed to reflect this revolutionary product. In addition, the company manufactured disc harrows, cultivators, chisel plows, and air seeders. Products were marketed throughout Canada and the USA.

Fair Grade Scales Ltd., Baytree, Alberta (James McDowell). Established in 1984, the company manufactured the "Fair Grade" scale and flow metre which was invented by James McDowell. Exports markets were established in Europe and the USA.

Falcan Industries, Fort Macleod, Alberta (James Vandervalk; Sheryl J. Vandervalk). Manufactured automatic fold-out cranes.

Faroex Ltd., Gimli, Manitoba (Betty Vigfusson; Perry Dankochik, Sales Mgr.). Established in 1980, the company manufactured floor grates and pens for livestock and fibreglass support frames for flooring. These products were invented by the company. Products have been exported to Australia, the EEC, Finland, and the USA.

Field Technologies (1989) Ltd., Saskatoon, Saskatchewan (Doug Barmby, Sales Mgr.). Established in 1987 after being a division of SED Systems since 1976, it was reorganized as a subsidiary of Flexi-Coil in 1989. The company manufactured air-seeder monitors, sprayer monitors and controls, and grain-loss monitors. It developed export markets in the USA.

Flexahopper Plastics Ltd., Lethbridge, Alberta (J.G. Spenceley, Pres.; J.W. Spenceley, Design Eng.). Products included grain hoppers, auger down spouts, chemical sprayer tanks, storage and nurse tanks, and livestock bunk feeders.

Flexi-Coil Ltd., Saskatoon, Saskatchewan (Terry Summach, Pres.; Dennis Summach, Vice-pres., Mktg.; Glen Pask, Vice-pres., Manuf.). Originally established by Emerson and Kenneth Summach in 1952 to manufacture a unique land packer, the company grew into a large multi-product firm. Ownership remained with the original families and products were all designed, manufactured, assembled, and marketed by the company. These included air seeders, tillage equipment, field sprayers, harrow packer drawbars, and packer drawbars. Goods were sold largely in western Canada, but were also exported to parts of Africa, Australia, eastern and western Europe, the Middle East, and the USA.

Forever Industries (1971) Ltd., Winnipeg, Manitoba (Ernie Janzen, Pres.; Otto Janzen, Mgr.; Bill Kinnear, Sec. Treas.). Herman Roecker established the company in the early 1920s to manufacture a grain cleaner of his own invention and patent. The company manufactured grain cleaners and other products including belt conveyors, scalpers, bucket elevator legs, and bag loaders. The company has found export markets in Africa, Australia, India, and the USA.

Forster Feeder Manufacturing Ltd., Killam, Alberta (Robert Forster; Joann Forster). Manufactured augers and swather transports.

Fouillard Trailer Sales, St. Lazare, Manitoba (Henry Fouillard). Manufactured stock and flatdeck trailers.

F.P. Bourgault Industries, Cultivator Division Ltd., and Air Seeder Division Ltd., both in St. Brieux, Saskatchewan (Gerald F. Bourgault, Gen Mgr.; Paul Leray, Vice-pres. & Prod. Mgr., Cult. Div.; Richard Coquet, Vice-pres. & Mfg. Mgr., Air Seeder Div.; Glenn Mark, Prod. Mgr., Air Seeder Div.; Alan Weenk, Sales Mgr.; Paul Boyer, Purch. Mgr.). F.P. Bourgault designed and manufactured a four-row cultivator in 1969 and incorporated his company in 1973. The company developed a full line of dual-purpose cultivators and continued to diversify. In 1985 it was formed into two divisions, and products included cultivators, chisel plows and other tillage equipment, wing sweeps, air seeders, sprayers, and packer bars. Export markets were established in the USA.

F.P. Bourgault Tillage Tools Ltd., St. Brieux, Saskatchewan (Rick Schemenauer; Stewart Hindley; Pat Yeager). Produced tillage tools including sweeps, spikes, chisels, and knives.

Franklin Enterprises Ltd., Winnipeg, Manitoba (Gus Beitz, Mktg.; Zoli Gergely, Sr. & Jr.). Manufactured "Bushel Master" grain vacs, displacement blowers, and cam lock couplings, and did custom machining and tool and die work.

Free Form Plastic Products Inc., St. Brieux, Saskatchewan (Glen Lefebvre; Gerald Lefebvre; Ronald Lefebvre). Manufactured auger spouts, hoppers, sprayer tanks, and air curtains, and did custom molding and plastic manufacturing.

Froc Industries Ltd., Melfort, Saskatchewan (Maurice Froc; Darin Froc; Perry Froc). Manufactured air seeders and seed boots.

Garratt Welding and Repair, Milestone, Saskatchewan (Lyle Garratt; Neil Garratt). Originally established in the early 1970s as an auto repair shop, the business gradually expanded to welding and light fabrication and began manufacturing bin cones, grain-cleaning equipment, and trailers, and did custom fabrication.

Gen Manufacturing Limited, Coaldale, Alberta (Henry Bergen, Pres.; David Bergen, Asst. Mgr.). Manufactured fertilizer banding knives, wear points for tillage equipment, and "easy lock" gate closures.

Geres Sales & Service, Neilburg, Saskatchewan (Peter Geres, owner; Lyle Thom, Off. Mgr.). Originally established in 1951 to manufacture rock diggers that were invented by Peter Geres. Gradually added other product lines including liquid supplement tanks and bale feeders.

Glendale Industries Limited, Brandon, Manitoba (Bill Hamilton; Glen Thomson). Established in 1984, the company initially manufactured liquid manure spreaders. The company grew as the product line gradually expanded to include equipment such as vacuum spreaders, and lagoon and pit pumps, all designed in-house.

Goldec International Equipment Inc., Red Deer, Alberta (Don Forsey). Manufactured "Nifty Feed" dispensers.

Grainhand, Div. 298157 Alta Ltd., Two Hills, Alberta (Henry Haag, Pres.; Joseph Galichowski, Vice-pres.). Manufactured grain-handling equipment.

Groening Industries Ltd., Treherne, Manitoba (Ben Groening). Product line included hitches, posthole diggers, mowers, track erasers, drill transports, and front-end loaders.

Hagen Electric (Lethbridge) Ltd., Lethbridge, Alberta (Mark Wheeler & Nick Trefan, Owners). Initially established in 1947 by Bernie Hagen, the company was then taken over and incorporated in 1962 by Allan D. Wheeler. Additional product lines initially included hand and power tools, and welders. Later, constant pressure control equipment and pump primers were added.

Harmon International Industries Inc., Saskatoon, Saskatchewan (Cal Moneo, Gen. Mgr.; Vic Moneo, Sales Mgr.). The company's products included air seeders, drawbars, field sprayers, harrows, tractor weights, and "Rock-o-Matic" rock removal equipment.

Harvest Services Corporation, Craik, Saskatchewan (Bert Wildfong, Pres.; Rick Wildfong, Sales Mgr.). Established in 1978 to rebuild concaves for combines. Later, the company manufactured new concaves and rub bars and gradually added other products such as adjustable sieves and chaffers and rear beaters, all of which were invented in-house. The company exported to Australia, France, and the USA.

Haukaas Manufacturing Limited, Mortlach, Saskatchewan (Greg Haukaas; Duane Haukaas). The company was established in 1980 to manufacture a swather-side draught kit that was invented and patented by Duane Haukaas. Later, other products were added including a ground-driven rod weeder attachment and side-arm field marker, both of which were invented in-house. The company exported to Australia and the USA.

High-Line Mfg. Inc., Vonda, Saskatchewan (Ray Bussiere; J.P. Hounjet; Gill Topping). In 1961 Rosaire Bussiere established a company called Rock-O-Matic and manufactured a rock picker of his own invention. The first plant was on his farm but he soon moved to the nearby town of Vonda. The company grew and was sold to Smith-Roles Ltd. of Saskatoon in 1978. In 1988 Ray Bussiere, son of the original owner, took over the plant and the company continued the production of rock pickers, as well as harrow and packer bars, sprayers, rock windrowers, and cultivators. The company also did steel fabricating. It exported to Australia, France, the USA, and the USSR.

Hi-Qual Mfg. Ltd., St. Boniface, Manitoba (E.A. Akins). Manufactured livestock-handling equipment.

Hi-Tech Fabricators Inc., Selkirk, Manitoba (Johann Sigurdson). Produced horse and stock trailers and industrial and specialized trailers.

Hold-On Industries Inc., Aberdeen, Saskatchewan (Floyd McMartin). Manufactured plastic products, hitches, potato diggers and planters, cultivators, and rototillers.

Holland Hitch Western Limited, Surrey, B.C. (Grant Jones, Gen. Mgr.; Matthew Fulton). Established in 1977, the company's major products were "Aer-Way" soil aerators and seedmatic seed drills. Export markets were developed in Mexico and the USA.

Honey Bee Manufacturing Ltd., Frontier, Saskatchewan (Gregory Honey, Pres.; Brad Nelson, Gen. Mgr.). In 1979 the owners incorporated the company to meet a demand for the rod weeder attachment invented by them and manufactured on their farm at Bracken. After developing a self-propelled swather designed to mount on a farm tractor, the company secured plant facilities and continued to expand its product line. The company exported to the USA.

Howard Concave Corp., Rosetown, Saskatchewan (Murray Howard, Pres.; John Howard, Sec. Treas.; Dale Skelton, shareholder). Established in 1984, the company manufactured concaves and repaired component parts for combines. It exported to the USA.

IBEC, International Business and Engineering Corporation, Spruce Grove, Alberta (Rosabelle Daugela, Mktg. Dir.; Mark Daugela, Eng. & Gen. Man.). Manufactured electronic/microprocessor control systems, grain-drying and conveying equipment, and alfalfa equipment.

Indal Metals, Saskatoon, Saskatchewan (G.B. Wenschlag; M. Maximuk). Originally incorporated as Westland Metals (Vancouver) in 1962, it was established in Saskatchewan in 1978. It manufactured agri roofing and siding and accessories, grain bins, and elevator siding, and did custom fabrication.

Inland Plastics Ltd., Drumheller, Alberta (John W. Goudy). Manufactured roll-up tarps and portable buildings.

Innovative Equipment Inc., Saskatoon, Saskatchewan (R. Barry Rogers, P. Eng., Pres.). Manufactured sprayers, fertilizer applicators, seeders, and custom-designed equipment.

Intercontinental Truck Body Ltd., Coaldale, Alberta (Jake Van Seters, Jr.). Manufactured aluminum grain boxes, cattle and poultry vans, beet end gates and trippers, horse trailers, and specialized van bodies.

International Fertilizer Systems Ltd., Calgary, Alberta (William F. Smith; Robert I. Martin). Manufactured fertilizer application, blending, and conveying components and accessories.

Inventronics, Moose Jaw, Saskatchewan (Mark Larocque). Manufactured farm power centres.

Jay-Lyne Products Mfg., Morden, Manitoba (Jake Guenther). Established in 1988, the company manufactured poultry-moving carts and a "Hold-o-Matic" holder used for moving poultry via overhead trolleys or conveyors. The company exported to Europe and the USA.

Jim Scharf Holdings Ltd., Perdue, Saskatchewan (Jim Scharf; Bruna Scharf). Established in 1985 and manufactured the "Safety-Hoe," a device for cleaning grain-truck corners.

Jodale Perry Corporation, Winkler, Manitoba (Rick Angers, Sales & Mktg.; Michael Reimer, Gen. Mgr.). Peter L. Reimer established the company in 1965 and it was later incorporated as Prairie Cab Ltd. It was reorganized as Jodale Perry Corporation in 1985. The original owner saw a need for a cab designed for western Canadian weather and the company continued to make ROPS all-steel tractor cabs as well as swathers and mowers. The company exported to Brazil, Taiwan, and the USA.

John Buhler Inc., Winnipeg, Manitoba (John Buhler, Pres.; Larry Schroeder, Gen. Sales Mgr.). Originally established in Morden by Adolf Krushel, who designed a grain crusher. Later, the company, Standard Gas Engine Works Ltd., also made augers and hammer mills. Krushel's son-in-law John Buhler took over management of the firm, and when he later bought the company, he changed the name to Farm King Ltd. Purchase of other companies in Winnipeg led to further name changes. Over the years, products included grain augers and cleaners, front-end loaders, harrow equipment, hydraulic cylinders, bale elevators, rotary cutters, and roller mills.

Jortar Industries Ltd., Moose Jaw, Saskatchewan (Ken Mews). Manufactured pick-up attachments for balers, forage harvesters, and other related products.

Kaletsch Rotor Corporation Ltd., Cluny, Alberta (Dr. R. Kaletsch; Ralf Kaletsch). Manufactured air reels, sprayer fans, packer wheel equalizers, rotor spacers, and stone kickers.

Keho Alta Products Ltd., Barons, Alberta (M. Roy Gullickson, Pres.; Zan Gullickson, Prd. Dev. Mgr.). Established in 1978 to manufacture a locally invented natural air drying system. Another major product, the windreel combine attachment, was added later. Other products included rod weeder attachments for cultivators and harrow attachments. The company exported to the USA.

Kellough Bros. Limited, Stettler, Alberta (Vic Kellough, Pres.; Hugh Kellough, Vice-pres.). Established initially in Camrose by Fred and Margaret Kellough in 1954 as a construction company. After moving the company to Stettler, Fred Kellough began designing and manufacturing discs suitable for the conditions and terrain of western Canada. The company and its product line expanded over the years and the "Kello-Bilt" tillage equipment included a wide range of implements. The company exported to Australia and the USA.

Kep Industries Ltd., Kerrobert, Saskatchewan (Erhard Poggemiller; Evelyn Poggemiller). Custom-manufactured electronic controls, and haying and cattle-feeding equipment.

Kerpan Welding Ltd., Watson, Saskatchewan (Raymond Kerpan, Pres.; Valerie Kerpan, Sec. Treas.). Established in 1972 and did custom manufacturing and repaired straw deflectors, tow bars, and trailers.

King of the Road Trailer Manufacturing Co. Ltd., Consul, Saskatchewan (Allen Moebis; Donna Moebis; Joan M. Moebis). Manufactured livestock trailers and associated agricultural equipment.

Kingsman Industries Div. of R-Co., Inc., Winnipeg, Manitoba (Randy Reyher; Bob Warburton). Produced automatic round bale movers with quarter-turn lift arms.

Kipp Kelly, Winnipeg, Manitoba (Brian Kelly; Dirk Moolman). Manufactured gravity separators, air flotation stoners, and rotary granulators.

Kirchner Machine Ltd., Lethbridge, Alberta (Marvin Kirchner, Vice-pres.; Jerry Kirchner, Sec.; Dwayne Kirchner, Sales Mgr.). First established in 1945 by brothers Harold and Earl Kirchner to market a haysweep and stacker designed by Earl. Later they expanded their product line to include land levellers and ditchers and set up a plant in Lethbridge. Many other implements were invented and manufactured over the years, including equipment for the sugar beet industry. The company exported products to the USA. In time, a second generation of brothers took over management of the company.

Koenders Mfg. Co. Ltd., Englefeld, Saskatchewan (Wilfred Koenders). Established in 1979 to manufacture implements invented by Wilf Koenders. These included windmill dugout aerators, bar fans and ladders, hog feeders, and barn heat exchangers. The company exported goods to the USA.

Kuelker's Mfg. Ltd., Didsbury, Alberta (Clem Kuelker, Pres.; Linda Buschert, Purch. & Off. Mgr.). Originally established in 1959 as a welding and repair shop. After a move to new premises, manufacturing began in 1971. The product line expanded and included silage and fertilizer wagons, manure pumps, bale handlers, and vacuum tanks. Goods were exported to the USA.

Kyle Welding & Machine Shop Ltd., Kyle, Saskatchewan (Ed Jacobsen; Richard Jacobsen). Ralph Jacobsen established the company in 1949 as a welding and machine shop. In 1955 he designed and began to manufacture water tanks. He also invented and patented a single disc field marker. Many other items were manufactured over the years, including disc sharpeners, swath guards, drop lifters, and bale shredders.

Labtronics, Winnipeg, Manitoba (William McRae; Dwight McRae). The company was established in 1972 to manufacture the model 919 moisture metre and was incorporated as Halross Instruments Corporation Limited in 1976. The metres were exported to Mexico and the USA, and the company supplied temperature probes and other related equipment to the food-processing industry and to farmers.

Lakesteel Industries, Shoal Lake, Manitoba (Derek Jenkins). Manufactured swing auger covers and utility augers, and did custom fabrication.

Laurier Manufacturing Limited, Vonda, Saskatchewan (Ray Bussiere; J.P. Hounjet; Gill Topping). Manufactured bale movers, harrow packer bars, and drill and swather transports.

Lea Action Axle (1989) Ltd., Rosetown, Saskatchewan (Robert Decker; Murray Button; Terry Sparks). Manufactured independent leaf spring suspension systems for trailers.

Leon Ram Enterprises Inc., Yorkton, Saskatchewan (Leon Malinowski, Pres.; Ray Malinowski, Exec. Vice-pres.). Established in 1952 by Leon Malinowski as a one-man enterprise. It expanded over the years and a wide variety of implements were manufactured, many of which were in-house inventions. Products included dozer blade units, front-end loaders, rear blades, rock pickers, scrapers, chisel plows, rod weeders, cultivators, and air seeders. The company exported to Australia, France, the USA, and the USSR.

Lethbridge Industries Ltd., Lethbridge, Alberta (Wayne Gayford). Manufactured roller chain sprockets and feed mills, and also offered machine shop services.

Lewis Cattle Oiler Co. Ltd., Oak Lake, Manitoba (Joe W. Lewis, Pres.; Garry R. Morcombe, Mgr.). Established in 1966 and manufactured cattle oilers, invented by Joe W. Lewis. The company enjoyed steady growth and developed export markets in the USA.

Lift Off Stock Trailers, Pilot Butte, Saskatchewan (David Farden; Pat Farden). Manufactured stock trailers and round bale movers.

Lode-King Industries, Winkler, Manitoba (Terry Elias, Pres.; Lloyd Elias, Sales Mgr.). Manufactured drill-fill systems, roll-over tarps, and hopper bottom grain trailers.

Loewen Mfg. Co. Ltd., Altona, Manitoba (Lloyd Loewen, Pres.; Warren Loewen, Gen. Mgr.). Established by Jake Loewen in 1959, it remained a wholly owned family business as the second generation entered the enterprise. Initially, the company produced bearing conversion kits, but it expanded its product line over the years to include combine replacement parts, field markers, grain drill recappers, and beet harvesters. Export markets were developed in Australia, Europe, Mexico, and the USA.

Loewen Welding & Mfg. Ltd., Matsqui, B.C. (Ernest Loewen; Wayne Raiche). Established in 1967 to work with individual farmers on their specific implement needs. This custom work continued but other implements were also manufactured, including manure spreaders, auger feed mixers, manure agitators, and conventional manure spreaders. Products were exported to the USA.

Lokken Ag Innovations, Weyburn, Saskatchewan (Ernest Lokken). Manufactured a cutting-height gauge for self-propelled combines.

Lonn Al Welding & Manufacturing Ltd., Bashaw, Alberta (Al Gerth; Lonni Gerth). Established in 1969 as Al's Welding, the company grew steadily and its expanded product line included bale stookers and forks, and dozer blades. It also did custom manufacturing.

M & R Machines Ltd., Weyburn, Saskatchewan (Brad J. Stewart). Established in 1972 by Sam Murray and acquired in 1984 by Brad Stewart, an employee of the firm. Major products included Hansen tow bars, discer packer hitches, discer dual hitches, and flatdeck trailers.

McCoy Bros. Group, Edmonton, Alberta (Terry McCoy, Pres.; Ron Kutney, Mgr., Axle Div.; Ian McCreath, Mgr., Spring Div.). Originally established in 1920. Manufactured cultivator shanks, coiled and leaf springs, rock picker tines, trailer axles, and components for grain-hauling equipment. Exported to the USA.

MacDon Industries Ltd., Winnipeg, Manitoba (Allan MacDonald, Pres.). Originally established in 1949 as Killbery Industries by Tom and Bill Killbery. It was acquired by Joe MacDonald and renamed MacDon Industries Ltd. in 1971. The company's products included pull-type and self-propelled grain and hay swathers, specialty crop headers, and combine headers, all designed and developed by the company. Implements were exported to Africa, Australia, China, Europe, South America, the USA, and the USSR.

Marshall Marking Systems Ltd., Turtleford, Saskatchewan (J.A. Marshall, Pres.; Brian Marshall, Sec.). Manufactured field markers, seed treaters, air model seed treaters, and stainless steel tanks.

Mass Load Technologies, Saskatoon, Saskatchewan (Bill Elliott, Pres.; Marcel Fendelet, Vice-pres.). Manufactured electronic weigh systems, feed wagons, hoppers, and cattle scales.

Mazurek Industries Ltd., Carrot River, Saskatchewan (Raymond Mazurek; Jean Mazurek). Manufactured tow bars, dozer blades, and trailers, and also did custom manufacturing.

Meldon Plastics Ltd., Regina, Saskatchewan (J. Donald Little; Melba P. Little). Manufactured fibreglass water and sewage tanks, F.R.P. pipe and flanges, pultruded rod and flat slat, and grating.

Melet Plastics Inc., Winnipeg, Manitoba (Ed Shinewald, Pres.; Carl Paziuk, Plant Mgr.). Produced mold-making and custom injection molding, and also plastic parts according to customers' designs.

Meridian Industries Ltd., Morris, Manitoba (Bernie Thiessen; Elmer Friesen; Ron Friesen).

Manufactured hopper bottom grain bins, silos, hopper tanks, and fuel tanks.

Metal Form 406884 Alta Ltd., Innisfail, Alberta (Russell Bennett; Brian Koole). The business was operated since 1974 by Alvin and Anne Hand and was incorporated as Metal Form in 1989. The company manufactured cabs, fuel tanks, feed system components, silage wagon flights, and chutes, most of which were manufactured for other companies.

Michel's Industries Ltd., St. Gregor, Saskatchewan (Teresa Eastman; Walter Michel). Manufactured side-rolling truck tarpaulins for grain boxes and semi-trailer units.

Midland Manufacturing Limited, Rosenort, Manitoba (Ervin Friesen, Pres.; Elmer Dueck, Gen. Mgr./Cont.). Established as a blacksmith shop in the early 1960s by Harry Isaac, the company was incorporated as Midland in 1976. It began manufacturing grain bodies in the early 1970s and later added roll tarps and farm end dump trailers. Products were exported to the USA.

MMP Spray Roller Co. Ltd., Swift Current, Saskatchewan (Morley Kolskog; Patricia Kolskog). Established in 1973 to manufacture MMP weed sprayers that were invented and patented by Morley and Patricia Kolskog. "Dial-A-Ball" hitches were later added to the product line.

Monarch Industries Limited, Winnipeg, Manitoba (Fred Preun, Prod. Mgr.; Bill Parkhill, Sales Mgr.). Manufactured hydraulic tie rod and welded cylinders, sprayer pumps, de-watering pumps, and liquid fluid fertilizer transfer pumps.

Morris Industries Ltd., Yorkton, Saskatchewan (Wendy Morris, Pres.). Originally established by George Morris in 1929 to market a rod weeder with a unique trip device. Following World War II, the company expanded and its product line eventually encompassed a wide variety of farm implements, including chisel plows, air seeders, floating hitch cultivators, packer harrow bars, press drills, harrow drawbars, and field cultivators. The company exported to Australia, France, Jordan, and the USA.

Morrow Industries Ltd., Lanigan, Saskatchewan (John Morrow, Pres.; Martha Morrow, Sec.). Manufactured stock trailers, flat-decks, and livestock-handling equipment, and performed custom fabrication.

Mumby Manufacturing Ltd., St Brieux, Saskatchewan (Dwayne Mumby; Judy Mumby). Established in 1984 to manufacture round bale transports designed by Dwayne Mumby. The company built and assembled the wagons in the farm shop and sold units to the Saskatchewan department of Agriculture and Food.

Munckhof Mfg., Co., Oliver, B.C. (Gerard Van Den Munckhof). Established in 1979 to manufacture weed and orchard sprayers, fork lifts, and irrigation booms, all of the company's own invention.

MWC Cradles Ltd., Vermilion, Alberta (Robin Selte). Manufactured calf cradles used for branding, vaccinating, and castrating.

Naicam Industries Ltd., Naicam, Saskatchewan (Pam Hofmann, Acting Gen. Mgr.; Meryl McGonigle, Plant Mgr.). Initially established in 1987 to manufacture fertilizer/grain storage bins and truck boxes. The company grew steadily, and tines and swath rollers, which it designed, were added to the product line. The company exported products to the USA.

Ne-De Systems Ltd., Riverhurst, Saskatchewan (Neal Moyse, Pres.; Dennis Obrien, Sec. Treas.). The company began in 1980 as a family owned and operated business. It designed and manufactured monitors for grain drill shafts and air seeders, and later added header height monitors. The company grew slowly from a local enterprise to one that exported goods to Australia, Europe, and the USA.

Neil's Welding and Manufacturing, Spalding, Saskatchewan (Neil Bourque, Owner/Operator). Manufactured spacers for seed drills, round bale feeders and forks, tree planters, and crop lifters.

New Noble Services Ltd., Nobleford, Alberta (Ward Krywolt, Pres.; Richard Albutt, Gen. Mgr.). Initially established in 1936 by Charles Noble to manufacture and market the

Noble Blade, an innovative product that cultivated the ground but left the stubble to hold the top soil in place. Following World War II, the company expanded considerably and developed and manufactured other tillage implements. In 1979 Versatile Corporation bought the company, and in 1987 a new group acquired the assets and renamed it New Noble. It manufactured some of the Noble products and, for a time, some of the Victory blades and air seeders. It exported products to the USA.

New-Way Irrigation Ltd., Taber, Alberta (Tom Paterson; Richard Hay). Manufactured irrigation systems, pumping units, pumping control systems, self-cleaning screen boxes, and pump primers.

Norbert's Manufacturing, Glenboro, Manitoba (Norbert Vanwynsberghe). Manufactured stock, ATV, and flatdeck trailers and bin-erecting cranes.

Nord-Sen Metal Industries Ltd., Winnipeg, Manitoba (J.M. Seniuk). Manufactured hydraulic truck hoists and bulk-handling equipment.

Notekeu Manufacturing, Mazenod, Saskatchewan (Dan Samoisette, Pres.). Established by Leo Mayge in Gravelbourg, acquired in 1985 by Glen Munford of McCord, and in 1990 by Dan Samoisette of Mazenod. The company manufactured squeeze chutes, corral panels, bale spears, farrowing crates, and round bale feeders.

NRW Manufacturing Inc., Riverton, Manitoba (Peter Reimer, Pres.). Established in 1986 to manufacture hopper bottom bins, bin cones, corral panels, and bale feeders.

Olson AG Innovations, Regina, Saskatchewan (Brian Olson). Manufactured power pins.

Pattison Bros. Agro Ltd., Lemberg, Saskatchewan (Rick Pattison, Pres.). Established in 1981 to manufacture liquid fertilizer wagons. With new facilities, the company expanded its product line to include liquid fertilizer distribution systems, tank trailers, and spoke wheel injection systems.

Paysen Livestock Equipment Inc., Central Butte, Saskatchewan (Brian Paysen). Manufactured livestock handling and feeding equipment.

Peacock Industries Inc., Saskatoon, Saskatchewan (Neil Wagner, Pres.; Herbert Urton, Vice-pres.). Manufactured granular applicators and broadcasters, chemical processing and packaging, aerosol and water base emulsions, and environmentally safe bran bait pesticide.

Penner Distributing, Altona, Manitoba (Melvin Penner; Hank Penner). Produced a chemical safety kit.

Phason, Steinbach, Manitoba (Ed Restau; Howard Klopak). Established in 1989 by Tony Friesen, the company produced a variety of temperature and environmental controls and electronic fencers, many of which were invented in-house. The company marketed throughout much of Canada and exported to the USA.

Phillips Shank Straightener Ltd., Weyburn, Saskatchewan (Larry Phillips). Established in 1985 to manufacture a portable shank straightener designed by Larry Phillips. This tool reshaped shanks while on the cultivator or chisel plow and, together with a portable press, was exported to the USA.

Plastic Rotational Kontainers Ltd., Saskatoon, Saskatchewan (Ken Lorman, Jr.; Larry Szeles). Established in 1989 to manufacture and assemble products such as slip and cylindrical tanks, feeders, grain handling hoppers, shuttle tanks, and shipping containers.

PM 2000 Preventative Maintenance, Humboldt, Saskatchewan (Dan Beuker; Russell Kemper). Manufactured energy-conserving lubricants and filters and performed oil and filter analysis.

Polyrama Plastics (1987) Ltd., Edmonton, Alberta (Jim Farnel, Pres. & Gen. Mgr.; Mario Skrypczak, Op. Mgr.). Manufactured storage tanks, custom molded tarps and covers, haystack covers, and irrigation lines.

Power-Matic Industries (1989) Ltd., Morris, Manitoba (Reinie Bobrowski, Gen. Mgr.). Began in 1970 as a welding and machine shop serving the local area. In 1977 Reinie and Herb Bobrowski designed and manufactured a harrow drawbar and additional new products included grain augers, bin unloading equipment, and packer bars. The company exported to the USA.

Prairie Farm & Ranch Supply, Teulon, Manitoba (Ralph Eichler; Chris Hurd). Manufactured livestock equipment including fencing, stock racks, and feeders. The company also did custom manufacturing.

Prairie Steel Products Ltd., Clavet, Saskatchewan (Lawrence Goebel). Manufactured culverts, well cribbing, and grain and fertilizer bins.

Prairie View Farm Equipment "76" Ltd., Assiniboia, Saskatchewan (Garth Delyea). Manufactured replaceable harrow tips.

Precision Metal Fabricating Ltd., Rosetown, Saskatchewan (Loren Katzenberger, Pres.; Alf Tide, Sales Mgr.). Manufactured "Rake-up" combine pickup rotary screens.

Precision Rubber Moldings Ltd., Ile des Chenes, Manitoba (Bernie Eckstein, Pres.; Ursula Eckstein, Sec.). The company began in 1974 in Winnipeg to serve the local market. Later, it expanded and acquired a plant at Ile des Chenes. Its products were all molded from rubber and phenolic and included air-seeder equipment, tarp straps, and metre rollers for fertilizer applicators. Many of these products were invented in-house.

R. Roth Enterprises International, Landis, Saskatchewan (Reg Roth). Manufactured the "Quick-N-Safe" chemical container rinser.

Ralph McKay (Canada) Ltd., Regina, Saskatchewan (Hans Gaastra, Pres.; Wayne Glushka, Sales Sup.). While established in Canada in 1976, the parent company in Australia began in 1932 when Ralph McKay began manufacturing disc blades and coulters. The Canadian company expanded and increased its capacity to manufacture tillage tools such as chisel plow sweeps, furrowers, and disc blades, and developed an export market in the USA.

Ram-Bull Ltd., Barons, Alberta (Geoff De Boer; Norm Ward). Manufactured cattle-handling equipment.

Redekop Chaff Systems Ltd., Saskatoon, Saskatchewan (Leo Redekop). Manufactured a complete line of chaff collection and handling equipment.

REM Manufacturing Ltd., Swift Current, Saskatchewan (Frank Rempel, Pres.; Bob Sonntag, Vice-pres.; Clarence Zacharias, Sales Mgr.). Began in 1966 with just one employee and manufactured a chaff blower. The company expanded its facilities several times, and developed new product lines, including agricultural springs, harrow tines, grain vacs, harrow carts, flexible harrow sections, rod weeders, and cultivator units. Frank Rempel designed most of the company's products.

Rite Way Mfg. Co. Ltd., Regina, Saskatchewan (Les Hulicsko, Pres.; Al Rodenbush, Sales Mgr.). Incorporated in 1970 and began manufacturing a rock picker in 1973. Its product line expanded over the years to include harrow/packer drawbars, rock windrowers, chisel plows, and drill transports, most of which were invented by Les Hulicsko. The company exported to Australia, Mexico, and Saudi Arabia.

Rodono Industries Ltd., Clive, Alberta (Jim Grose; Faye Grose). Established in 1987 to manufacture a hydraulic header reverser for combines, which was invented by Jim Grose. This product, together with hydraulic vent openers and truck box latches, was manufactured in a facility on the owner's farm.

Rogers Engineering Inc., Saskatoon, Saskatchewan (Barry Rogers). Began in 1982 on a farm and moved into Saskatoon the next year. The company concentrated on research and development of the Windfoil sprayer and fertilizer injection equipment, and developed export markets in Britain, Germany, and the USA. Later became Innovative Equipment Inc.

Royal Trailer Co. Ltd., Lethbridge, Alberta (Roy Androkovich; Inga Androkovich). Established in 1978 as a one-person operation and gradually expanded over the years. The company manufactured horse and stock trailers in both aluminum and steel, and also flatdeck and gooseneck trailers, all designed by Roy Androkovich. It exported to the USA.

Rudelich Contracting Ltd., Picture Butte, Alberta (John Rudelich; Jim Rudelich; Doug Rudelich). Manufactured grain scalpers, trailers, augers, and bale forks, and also did custom manufacturing of farm equipment.

St. Albert Welding Co. Ltd., St. Albert, Alberta (Ed Lemiski). Manufactured trailers.

Sakundiak Equipment Ltd., Regina, Saskatchewan (Peter Sakundiak, Pres & Gen. Mgr.; Larry Sakundiak, Vice-pres. & Asst. Mgr.). Established in 1948 in a small plant in Regina to construct a disc harrow designed by Alex Sakundiak, brother of the company's president. Shortly after, the firm began to manufacture grain augers and remained with this product, gradually offering an increasing number of ever-larger sizes. The company exported to the USA.

Scarrow Industries, Spy Hill, Saskatchewan (Dennis Scarrow; William Gilchuck). Established in 1976 and remained on the farm where business started. It increased its product lines and number of employees and manufactured corral panels, bale feeders, trailers, and livestock handling and feeding equipment, and did custom manufacturing. It exported to the USA.

Schmidt Manufacturing Inc., Maple Creek, Saskatchewan (Arnold Schmidt; Pearl Schmidt). Manufactured flour mill baggers.

Schuh Distributors Ltd., Dilke, Saskatchewan (Stan Tait). Established in 1961 by Jack Schuh and manufactured a Watermaster pump. Later, Stan Tait invented and patented the "Turb-O-K-Vac" grain vacuum and developed an export market in the USA.

Schulte Industries Ltd., Englefeld, Saskatchewan (Earl R. Schulte, Vice-pres. & Mgr.; Jim A. Carnago, Mktg. Mgr.). Evolved from a blacksmith business initially established in 1914. In 1932 John Schulte joined his father in the business and expanded it to include repair of farm machinery. Later, the company began to design and manufacture various farm implements, including brush cutters, rock pickers, rock windrowers, cultivators, and rotary cutters. Export markets included Australia, Italy, Saudi Arabia, Syria, and the USA.

Senstek Ltd., Saskatoon, Saskatchewan (W.P. [Bill] Strelioff, Gen. Mgr.; R.M. Ball, Op. Mgr.). Established in 1976 and manufactured electronic weighing equipment, tillage-depth and header-height control systems, and shaft monitors. Most products were developed by Bill Strelioff and S.K. Wu, and export markets included Australia and the USA.

Setter Manufacturing Division, Russell, Manitoba (Greg Setter, Pres.; Cathy Setter, Sec. Treas.). Manufactured sprayers and spraying equipment, high power washers, and spot sprayers.

Shape Industries Inc., Winnipeg, Manitoba (Ryan Smith, Pres.). Manufactured grain bin ladders, hydraulic tube assemblies, and other agricultural parts and components.

Shur-Lift Industries Ltd., Winnipeg, Manitoba (Todd Knight, Sales & Service; Tom Achenbach, Vice-pres., Mfg.). Manufactured hydraulic cylinders, implement jacks, high pressure washers, cranes, scissor lifts, and hydraulic presses.

Shurtight Mfg. Ltd., Forestburg, Alberta (Merlin Badry; Louise Badry). Manufactured spacers for seed drill castor wheel assemblies.

Sickle Master Limited, Khedive, Saskatchewan (John M. Froh, Pres.; Bob Froh, Treas.; Les Froh, Sec.). Established in 1985 to manufacture the "Sickle-master," invented and patented by John Froh. The machine hydraulically removed and replaced the sickle section on a cutting bar.

Skyway Grain Systems Inc., Airdrie, Alberta (Ellwood J. Sawby). Products included grain-handling systems, pump-loop systems, elevator legs, swing arms, dual augers, hopper bins, and flow dryers.

Slimline Manufacturing Ltd., Penticton, B.C. (Kim Blagborne; Terry Blagborne; Bob Blagborne). Manufactured agricultural and industrial sprayers.

Somervillesteel Industries, Endiang, Alberta (Bill Somerville). Manufactured livestock equipment.

Spierco Industries Ltd., Calgary, Alberta (W.A. Spiers, Pres.). Established in 1971, the company manufactured granular applicators for fertilizers, herbicides, and insecticides. Export markets were Australia and the USA.

Spray-Air Canada Ltd., Carseland, Alberta (Barry Monner; Arne Johansen). Established in 1983 to manufacture air-assisted field sprayers. In 1988, a mechanically driven swing auger, invented by Barry Monner, was added to the product line. The company exported to the USA.

Spray-Tech-Systems Ltd., Vonda, Saskatchewan (Aime Bussiere; Laurent Bussiere). Product lines included estate and all-terrain vehicles, spot sprayers, and hitch sprayers.

Springland Manufacturing, Rivers, Manitoba (John Lepp; Henry Lepp; Herman Lepp). The company manufactured unloading equipment for farm and commercial bins.

Standen's Limited, Calgary, Alberta (Dieter Schwarz, Parts & East Sales Mgr.; Mel H. Svendsen, Vice-pres. & Gen. Mgr.). Cyril Standen incorporated the company in 1924 to manufacture car springs. It grew steadily, particularly in the 1970s and 1980s. The company was instrumental in developing the "Edge-On Forged Cultivator Shank" and its manufacturing line included leaf springs for all sizes of vehicles, and grouser bars for terrain-tracked vehicles. Export markets were in Australia and the USA.

Star Industries Ltd., Gravelbourg, Saskatchewan (Ernest Aussant, Pres.; Pauline Aussant, Sec.). Incorporated in 1974, the company manufactured land levellers and recaps for press drill packer wheels. The product line expanded later to include livestock equipment, corral panels, and round bale feeders.

Stirdon Systems Ltd., Olds, Alberta (Stirling McLeod; Allan Halverson). Manufactured overhead storage units, Iserex feed jets, electric feed gates, and Berex blowers.

Sunnybrook Welding and Machine Shop, Sunnybrook, Alberta (Audun Njarheim; Evelyn Njarheim). Manufactured combine concaves and various improvement parts for combines, rub bars, back beaters, and eco bait stations.

Sunset Solar Systems Ltd., Assiniboia, Saskatchewan (Doug Cameron, Owner/Mgr.; Lorraine Cameron, Owner/Sec.). Established in 1983 in a facility on the owner's acreage. Within two years, sales were such that the company moved into a plant in Assiniboia. It manufactured "Little River" pond mills for aerating dugouts, and "Row Hoe" tree weeders for cultivating shelter-belts. Both products were invented by Doug Cameron. Export markets were in the USA.

"Super Blow Gun," Melfort, Saskatchewan (Charlie Hopkins). Manufactured the "Super Blow Gun."

Super Trail Manufacturing, St. Alphonse, Manitoba (Lawrence G. Delichte; Yvonne G. Delichte). Manufactured bale feeders for large round bales.

Superior Machine and Welding, Carroll, Manitoba (Greg McFarland). Involved in custom fabrication and repair and prototype development.

Supreme International, a Division of Alteen Distributors Ltd., Wetaskiwin, Alberta (George Hunerfauth; Peter Hunerfauth). Established in 1953 as Alberta Engineering Ltd., a distributing company, and began by adapting their products for local needs. In 1957 the company started manufacturing its own lines, including farm gates and bale carriers, later adding V-rake trailers, and, among others, a cutter mixer feeder.

Swift Machine & Welding Ltd., Swift Current, Saskatchewan (Ken Dyck, Pres.). Manufactured swather stabilizers and offered general machining and welding services.

Symons Metal Workers Co. Ltd., Rocanville, Saskatchewan (Edna Rushton). Manufactured hand pump oilers and a garden tool line.

Tait's Grain Vacuum Systems Ltd., Dilke, Saskatchewan (Greg Tait; Stan Tait; Terry Scot). Produced floating water pumps.

Talon Tool Corp., Frontier, Saskatchewan (Olaf Friggstad; Howard J. Groth). Manufactured industrial and agricultural equipment.

Tansley Manufacturing Inc., Tourond, Manitoba (Pat O'Malley, Pres.; Jeff Lord, Sec.). Established in 1977 as a one-person business doing machinery repairs, the company gradually developed into a manufacturing concern. All of its products were invented in-house. The product line included grain probes, augers, vacuum pump blades, and pit-cleaning systems. Some exports were made to the USA.

Taylor Industries Ltd., Melfort, Saskatchewan (Stan Larson; Ernie Sorokowski). Manufactured fertilizer and grain bins, fuel tanks, swath rollers, and cement silos.

Terra Power Tractor Co. Ltd., Saskatoon, Saskatchewan (Liviu Dragomir; Olie Riffel). Established in 1973, the company assembled Universal tractors, combines, and other farm implements, mainly imported from Romania. Export markets were in the USA.

Texall Industries, Blaine Lake, Saskatchewan (Ken Sawitsky). Manufactured Texas gates, corral panels, and portable cattle chutes.

Thor Manufacturing Ltd., Winnipeg, Manitoba (Al Rudy, Pres.; Jim W. Kleebaum, Sales Mgr.). Manufactured "Conveyair" grain vacs, air locks, and cam lock couplers, and offered production machine shop services.

Traders Tec-Systems Controls Ltd., Bowden, Alberta (Dave Siemens, Gen. Mgr.; Russell Hunt, Op. Mgr.). Manufactured custom control panels for automating grain handling, feed processing, and seed cleaning systems.

Trailers Unlimited Inc., Edmonton, Alberta (Lorne From; Judy From). Established in 1987 to build small utility trailers designed by Lorne From. The company gradually increased the number and size of trailers constructed and in 1990 developed a new fibreglass-covered trailer.

Trailking Ltd., Brandon, Manitoba (Andrew Mills; Wayne Gurr). Manufactured livestock trailers, inspected and repaired trailers, and did custom metal fabricating.

Trail-Rite Flatdecks Ltd., Tisdale, Saskatchewan (Ray Peterson, Gen. Mgr.; Ken Magnus, Sales Mgr.; David Burton). Established in 1975, the company diversified its product line to include hopper bottom bins, modular bins, conversion cones, fuel storage tanks, platforms, ladders, cages, and towers. The company exported 10 percent of its production to the USA.

Trimeg Holdings Ltd., Calgary, Alberta (G. Shackleton; B. Cooke). Manufactured flexible bulk containers.

Tri-Met Instruments Ltd., Winnipeg, Manitoba (Alan M. Roulston). Produced a moisture meter for grain and some mining equipment.

Tube-O-Later Inc., Thorsby, Alberta (Gerry Brodrecht). Manufactured round bale baggers with a grain attachment, and automatic round bale wagons.

Uniflyte Co. Ltd., Winnipeg, Manitoba (Wilf Piche, Gen. Mgr. & Vice-pres.; Gerry Cadorath, Vice-pres., Mktg.). Established in 1975, it manufactured auger flighting, and angle and flat rings that were marketed across Canada and the mid USA. The angle and flat rings were designed in-house.

Univision Industries Ltd., Humboldt, Saskatchewan (Ken Esler, Pres.; Gordon Lehne, Gen. Mgr.). Established originally in Biggar in 1969 and later expanded to Humboldt. Began with the manufacture of small trailers and farm truck boxes, then moved into semi-trailers and grain bins, most of which were designed by Ken Esler. Products were marketed across Canada and into the USA.

Urvold Industries Ltd., Nobleford, Alberta (Harold Urvold). Manufactured straw and chaff blowers, swath attachments, packer wheel caps, and angle iron retreads.

Valcon Equipment—Division of Vale Farms Ltd., Indian Head, Saskatchewan (Jim Halford; Larry Koturbash; Ken Drever). Manufactured the "Conservo Pak Seeder" and "Poly Packer Gangs."

Valmar Airflo Inc., Elie, Manitoba (Gary Dunbar; Jim Gladstone, Sales & Mktg.; Ray Rivard, Purch.). Established in 1977 to manufacture implements designed by its principal owner, Charles Balmer. The company grew steadily and products included granular herbicide applicators and fertilizer spreaders and were marketed in the western provinces and mid-west USA.

Vansco Electronics Ltd., Winnipeg, Manitoba (Ed Van Humbeck, Pres.; Gino Kost, Mktg. Rep.). Manufactured monitoring and control systems, air-seeder monitors and blockage sensors, and actuators.

VFEO–Ford New Holland Canada Ltd., Winnipeg, Manitoba (Paul M. Soubry, Vice-pres. & Gen. Mgr.; Reginald F. Walsh, Mgr., Intl. Sales; Major Rodd, Contr.). The origins of this company stretch back to 1945 when Peter Pakosh designed and manufactured a grain auger in the basement of his house in Toronto. After he designed a field sprayer, he and his brother-in-law Roy Robinson rented space and formed a partnership. In 1951 they moved to Winnipeg to be closer to their primary market. Considerable expansion took place, and in 1963 it became a public company, renamed Versatile Manufacturing Ltd. After the two principals retired, the company merged with Cornat Industries, Vancouver, and acquired Noble Cultivators of Nobleford, Alberta. Later, Ford New Holland took over the company, and in 1991, Fiat, a major European company, acquired 80 percent of the US-based Ford New Holland. It manufactured four wheel drive and bidirectional tractors, self-propelled and push swathers, pull-type combines, loaders, and backhoes.

Vicon Mfg. Inc., Portage la Prairie, Manitoba (Chuck Balmer, Sales & Mktg.; Doug Pedden, Finance). Manufactured chisel plows, field cultivators, discers, air seeders, harrows, swathers, combines, round balers, sprayers, and Noble blades and drills.

Victory Equipment (1983) Ltd., Lethbridge, Alberta (Ray Dyken, Pres.). Victor Erdman, who worked for Noble Cultivators, designed a V-shaped blade and started his own company in 1940. Gradually other product lines were added, including harrows, air seeders, hydra rods, and windrow pick-ups.

Vonda Welding & Machine, Vonda, Saskatchewan (George Quaroni). Manufactured bale wagons, and did custom manufacturing.

W. Goertzen Trailer Ltd., Airdrie, Alberta (Walter Goertzen; Mike Goertzen; Roger Goertzen). Manufactured livestock and horse trailers, and utility and flatdeck trailers.

Wabash Welding Ltd., Westlock, Alberta (Walter Kostiw; Valerie Kostiw). Established in 1980 in a shop on an acreage. Originally worked mainly for the oil industry, but later concentrated on the farming industry. The company moved into Westlock where it manufactured land rollers, bale wagons, rake transports, bale forks, and drill recaps.

Wahpeton Canvas Co. Canada Ltd., Avonlea, Saskatchewan (Thomas D. Shorma, Dir. Mktg.; Joyce Wagner, Branch Mgr.). Manufactured roll-up, fitted, and flat tarps, side kits for flatbed trailers, and tarp repair kits.

Weninger Industries, Saskatoon, Saskatchewan (Ross McDonald, Op. Mgr.; Brant Regush, Sales Mgr.). Ozzie Weninger began the company as a welding shop in 1974, where he did general repair and small fabrication. The following year he started to manufacture and the firm enjoyed steady growth. Products included bin cleaners, bale feeders, drill fills, hopper bottom grain and fertilizer bins, and retrofit hopper bottoms. In 1989 the company was sold to Jannock Steel Fabricating Co. Ltd.

Westeel Agricultural Products, Winnipeg, Manitoba (Don Taylor, Gen. Mgr.; Harry Harms, Western Sales Mgr.). Winnipeg Ceiling and Roofing Co. was established in 1905, and as steel began to be used for many more products, the name was changed to Western Steel Products Limited in 1920. The company acquired many other plants across the country and following World War II changed its name to Westeel Products Limited. In 1987 Westeel design engineers designed "The Boot," an aeration system, steel foundation, and hopper assembly for use with grain bins. Jannock Limited became a major shareholder and after a partnership arrangement with Stelco, the company became known as Westeel—a Division of Jannock Steel Fabricating Co.

Western Combine Corporation, Portage la Prairie, Manitoba (Frank Pickersgill, Pres.; Ralph Fairbairn, Contr.). Established in 1989 to assemble a self-propelled combine harvester (rotary-axial flow), the product rights for which were obtained from the Massey Combine Corporation. Component parts came from Ontario, other regions of Manitoba, and the

USA. The company marketed the product across Canada and to the USA.

Western Concord Mfg., Edmonton, Alberta (Mark Frohlich; Vern Mudryk; Lyle Hamilton). Manufactured agricultural film, poly bags, garbage bags, shrink film, pallet covers, box liners, blast hole liners, and silage film.

Western Industries (St. Gregor) Ltd., St. Gregor, Saskatchewan (Joe Menz, Pres.; Pat Blanch, Sec. Treas.). Manufactured grain boxes, farm wagons, fertilizer tanks, drill fills, and seed treaters.

Westfield Industries Ltd., Rosenort, Manitoba (Frank Plett, Pres.; Stan Plett, Sales Mgr.). Established in 1950 to manufacture augers, the company grew considerably and established branches in Saskatoon and West Fargo, North Dakota. The original owner's four sons bought the business in 1965, redesigned the product line, and added tractor-operated and utility augers, and drill fills. Products were exported to Australia, Germany, Mexico, Saudi Arabia, and the USA.

Westland Plastics Ltd., Winnipeg, Manitoba (Tom Wilton, Pres.; Steve Bonner, Sales Mgr.). Established in 1981, the company manufactured sewage basins, ventilation fans, hog feeders, and hopper tanks, all designed in-house by Ivan and Steve Bonner and Brian Parker. In 1986 the company purchased Prairie Pride Enterprises and T & L Distributor.

Westside Enterprises, St. Brieux, Saskatchewan (Frank Nagy, Owner; Lenard Horwath, Sales Rep.). Manufactured all-season trailers and nagy buggies.

Wheatheart Hydrostatic and Machine Ltd., Saskatoon, Saskatchewan (David Quick, Foreign Sales Mgr.; Bob Quick, Gen. Mgr.). The company began with the manufacture of bin sweeps on the Quick farms in 1975. In 1981, Bob, Jim, and David Quick and Al McConnell incorporated Wheatlands, and established it in Saskatoon where it designed and manufactured an expanded product line, including self-propelled auger transports, post hole augers, semi-transfer augers, and the "Foot Saver," a safety auger assembly. The company exported to Australia and the USA.

Wheatland Bins Ltd., Lethbridge, Alberta (Menno Kornelsen, Pres.; Richard Friesen, Sec.). The company manufactured hopper bottom fertilizer and grain storage bins that were first designed by Elmer and Ron Friesen in 1965. They were with Meridian Industries in Manitoba at that time but later became partners in Wheatland. Within two years, the company expanded its premises and in 1986 opened a second facility in Camrose, Alberta.

Wilger Industries Ltd., Saskatoon, Saskatchewan (Wilfred H. Wilger, Pres.). Established in 1975, the company began with the manufacture of farm field sprayers invented by Wilf Wilger. A variety of different models was subsequently produced as well as sprayer nozzle bodies. The company exported to the USA.

Winn-O-Bar Enterprises Ltd., Newdale, Manitoba (Cliff Usick, Pres.; Randy Usick, Sec. Treas.). Cliff Usick established the company in 1986 when he patented a combine shoe attachment. Later, he invented and manufactured a combine walker attachment. Markets included exports to the USA.

Zeebest Plastics of Canada Inc., Edmonton, Alberta (Derek Nielsen, Pres.). Manufactured holding tanks, sprayer tanks, barrels, livestock tanks, and feed troughs.

Index

Note: A number in bold face indicates a photograph.

Index